D1548141

IN-CIRCUIT TESTING

IN-CIRCUIT TESTING

John Bateson

VNR VAN NOSTRAND REINHOLD COMPANY
New York

Library of Congress Catalog Card Number: 84-27128
ISBN: 0-442-21284-4

Manufactured in the United States of America

Published by Van Nostrand Reinhold Company Inc.
135 West 50th Street
New York, New York 10020

Van Nostrand Reinhold Company Limited
Molly Millars Lane
Wokingham, Berkshire RG11 2PY, England

Van Nostrand Reinhold
480 Latrobe Street
Melbourne, Victoria 3000, Australia

Macmillan of Canada
Division of Gage Publishing Limited
164 Commander Boulevard
Agincourt, Ontario M1S 3C7, Canada

15 14 13 12 11 10 9 8 7 6 5 4 3 2 1

Library of Congress Cataloging in Publication Data

Bateson, John.
In-circuit testing.

Bibliography: p.
Includes index.
1. Printed circuits–Testing. I. Title.
TK7868.P7B37 1985   621.381'74'0287   84-27128
ISBN 0-442-21284-4

To my wife, JoAnn

PREFACE

The aim of this text is to increase your understanding of the methods employed for improving the quality of printed circuit boards (PCBs) in a practical manufacturing environment, by discussing printed circuit board faults and the test strategies implemented to detect these faults. This text emphasizes in-circuit testing as a prime test and diagnostic technique.

Test strategies are described — implementing functional board testers, in-circuit board testers, in-circuit analyzers, and loaded-board shorts testers. Also discussed are in-circuit tester's hardware, software, fixturing, and programming. Specific attention has been given to the in-circuit tester's capabilities and limitations, features and benefits, advantages and disadvantages. Chapter 5, as part of the total production testing process, discusses rework stations, networking, and test area management. Chapter 8 is devoted to discussing the benefits derived by employing in-circuit testing in the service repair arena. This text concludes with chapters on vendor investigation and a financial justification.

Additional emphasis is placed on having design engineering acquire an interest in manufacturability, testability, and the importance of consulting with manufacturing early in the design process. This book is designed for ease of reading and comprehension for all levels of interest: ATE students, first-time ATE users, as well as those involved in test, manufacturing, quality control or assurance, production, engineering, and management.

JOHN BATESON

ACKNOWLEDGMENTS

I would like to express my appreciation to my colleagues: Donald Lenhert, Ray Young, Martin Ostrego, Tom Bush, and Greg Geary who technically reviewed this manuscript to ensure to the best of their knowledge, an accurate portrayal of the facts. Further, I would like to thank Maggie Hilligoss who typed this manuscript in a most diligent and conscientious manner.

I wish to express my appreciation to Factron Schlumberger for total access to the production board testing files which provided the data base from which this text was written. Also my appreciation to Prime Data as the source of a large amount of the statistical data incorporated in this text.

INTRODUCTION

The mission of electronic production is to optimize product quality and establish acceptable throughput or to optimize throughput and establish acceptable product quality, both in the most cost effective manner possible. Our emphasis will be on optimizing throughput with acceptable product quality, because this is the production philosophy in most commercial electronic companies.

First it is essential to select a test strategy and choose the proper test equipment to implement that strategy. Both the test strategy and the equipment selected to implement that strategy are functions of four manufacturing process variables: first pass yield, production volume, fault spectrum, and board mix. Therefore, the cost performance ratio of the selected test equipment will vary with the manufacturing process variables used to achieve the test strategy. That is, cost is a quantitative measurement, while performance, although definable, is more subjective and qualitative.

An effective test strategy should evolve concurrently with a company's three-year or five-year business plan. Management must systematically implement the long-range test strategy as a function of the financial justification for additional test capability based largely on cost reduction. Financial justification is more rewarding, however, when it is based on cost avoidance. Critical decisions must be made on how to expand a manufacturing test capacity within the scope of the budget allotments.

Why employ automatic test equipment for PCB production testing? The answer is, increased profit. ATE can help you increase productivity; decrease production costs; reduce requirements for skilled labor; and increase product reliability. The cost of detecting and reworking a PCB fault increases significantly as the production process advances and the product is shipped. An acceptable rule of thumb states the cost of detecting and correcting a fault increases tenfold for each successive stage in the production process. Further, the cost of repairing a product in the field increases an additional tenfold. The cost of finding a defective component at incoming inspection is between 30 and 65 cents. The cost of correcting a fault after PCB assembly and soldering ranges from $3.99 to $6.50. The cost of correcting a fault at system test ranges from $30.00 to $65.00. Further, correcting a defect in the field costs between $300.00 and $650.00. Obviously, it is economically advantageous to find and rework faults at the earliest possible point in the production process.

There are four general-purpose types of commercially available automatic PCB test equipment to be configured to implement a particular test strategy: loaded-board shorts tester (LBS), in-circuit analyzer (ICA), in-circuit tester (ICT), functional board tester (FBT). Each have certain advantages and disadvantages. A sound test strategy configures these test systems and in-house built test systems so that the strengths of the individual test systems compliment each other to provide an optimum solution to the production test requirements. In many cases, a series of test system configurations are employed at various stages in the manufacturing process development to fine tune the process in an effort to optimize the yield and product throughput.

The selection of automatic test equipment is a business decision. The three prime factors in your decision as illustrated in Fig. 1, are technical, economical, and risk. This text addresses in-circuit testing philosophy, architecture, software, and its effective use in production and service, keeping in mind the three prime business decision factors. When your manufacturing capacity approaches 100 percent, there is a total loss of flexibility to respond to higher levels of product demand. The backlog builds to an excessive size and the market growth is limited. The question is what test strategy do you employ to relieve the problem? When your service department has an enormous board float and there are shelves full of field returns plus excessive spares inventory that is still growing, again the question is what test stragety do you employ to relieve the problem? This book presents some guidelines, tools, and food for thought, to be used when you analyze your production and service testing requirements.

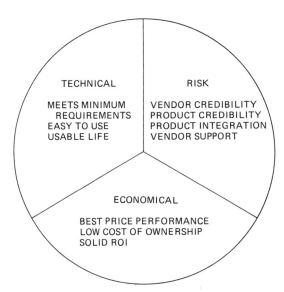

Fig. 1. Business decision.

Because of the large number of variables, general solutions cannot be declared or considered valid for all cases. Therefore, after careful study and analysis, the correct test strategy for your production or service requirement is your decision.

CONTENTS

IN-CIRCUIT TESTING

1
UNDERSTANDING THE PRODUCTION
FAULT SOURCES

Faults on loaded printed circuit boards (PCBs) are divided into three categories: (1) component faults, caused by components failing to meet specifications; (2) manufacturing faults, due to errors encountered in PCB assembly and/or soldering; and (3) performance faults, generally resulting from weak design or dynamic device failure. Fig. 1.1 illustrates a typical production flow and indicates the sources and potential magnitude of PCB faults.

1.1. COMPONENT FAULT SOURCES

The production manager is first interested in the quality of the parts he receives from raw material inventory. Between 10 and 20 percent of the faults at the PCB level are potentially caused by defective components. Do you purchase tested components or perform incoming testing, and to what degree? Do you perform 100 percent testing of all components, or semiconductor devices only? Are all components tested against an acceptable quality level (AQL)? The fact that a component's quality and reliability is the sole responsibility of the supplier does not help an electronics manufacturer when his product has a component failure. A company's incoming test philosophy appears to be based on the company's end product liability, the perception of the supplier's quality, and the economics of incoming test. Without question, when the end product involves human life or is in an environment where repair cannot realistically be performed, in space for example, 100 percent incoming test followed by testing at each phase of manufacturing is essential. Otherwise, it is more of a business decision. If a company believes a supplier is reliable and dependable, and the end product does not carry any major liability, an audit/monitor incoming function may be all that is required. However, we all have a tendency to generalize

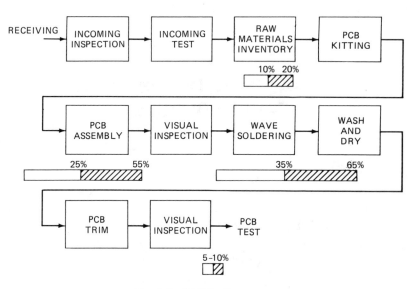

Fig. 1.1. PCB fault sources.

and not all components are treated the same. Component supply and demand cause large fluctuations in parts prices, premiums as high as six times normal list price. Who is to say that the part's quality is not affected by fast time-to-market demands. Further some parts are more critical in a product than others. Therefore, it is not unreasonable to find companies who perform 100 percent incoming test on some parts while performing an AQL on others and not testing others.

It may be true that today customer demands for high quality and reliability forces suppliers to perform more extensive characterization and reliability studies than in previous years. New components on the market cause suppliers to spend millions of dollars developing the capability of testing before shipping the product to their customers. Companies today cannot afford the complex test equipment and skilled personnel necessary to verify the supplier's parts or the quantity of defects does not justify incoming test recurring costs. All these statements may be true.

A Prime Data survey conducted in late 1983 indicated approximately 29 percent of the industry performs 100 percent incoming test on VLSI, 27 percent on LSI, 24 percent on SSI/MSI and 22 percent on analog components. Further, 27 percent of the industry

TABLE 1.1. Typical PCB Distribution for 175-Component Board.

Faults per PCB	Component Defect Rate		
	0.5%	1%	2%
0	41.8%	17.8%	3.3%
1	36.3	30.4	11.3
2	15.9	26.2	18.6
3	4.7	15.4	21.1
4	1.1	6.9	18.4
5+	0.2%	3.4%	27.2%

performed incoming AQL testing on VLSI, 38 percent on LSI, 45 percent on SSI/MSI, and 50 percent on analog components. This result leaves a large portion of the industry that does not perform incoming testing of components. Mathematically, the distribution of defective components will follow a binomial distribution when kitted for printed circuit boards assembly, resulting in an exponential failure rate.

Table 1.1 illustrates the fault distribution for three rates of component defects on boards with 175 components. Note that the first-pass-yield responds exponentially when the reject rate is halved. On average, for a 1-percent component defect rate, the first-pass-yield is 17.8 percent; 30.4 percent of the boards will have one defective component per board; 26.2 percent of the boards will have two defective components per board; and 15.4 percent of the boards will have three defective components per board.

Figure 1.2 illustrates the impact of installing defective components on a PCB. For a 1-percent defect rate, the magnitude of the failures increases exponentially with the component population. Note that the first-pass-yield for a 50-component board is slightly in excess of 60 percent, while it is approximately 2 percent for a 400-component board. The binomial distribution of multiple faults is also illustrated where the number of single faults per board decreases dramatically as the component population increases. The quantity of multiple faults per board increases exponentially.

Unless a component is extremely mature, testing to determine if it is good or bad does not ensure its reliability. Fig. 1.3 illustrates the failure distribution of semiconductors as a function of operating time.

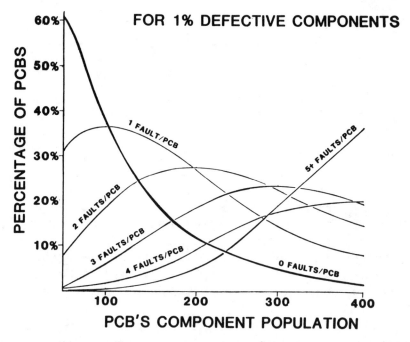

Fig. 1.2. PCB fault distribution.

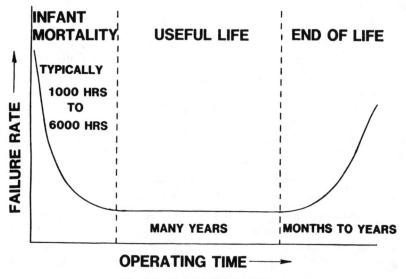

Fig. 1.3. Failure distribution.

The "bathtub" failure distribution curve begins in the infant mortality region typically ranging from 1000 to 6000 hours. Statistically, after approximately 250 hours of operation, 60 percent of the unstable devices will have failed. After approximately 450 hours, 90 percent of the unstable devices will have failed. The remaining 10 percent of the devices may take hours, days, weeks, months, or even years to fail.

To detect earlier life failures, many companies employ stress testing. Power-up/burn-in at high temperatures is a common method of stress testing. Nominally, the operation time doubles for every 10 percent above normal operating junction temperature. The most common burn-in setting, based on industry statistics, is approximately 125 degrees Fahrenheit for a period of 12–24 hours.

This type of stress testing will eliminate the majority of unstable devices before they are inserted on the PCB, especially those subject to infant mortality failures. Theoretically, testing 100 percent of incoming components should result in a PCB fault source of approximately 2–10 percent. When 100 percent component test is combined with 100 percent stress testing, the PCB fault source should be reduced to between 0 and 5 percent. Variations in these extremities will result in a proportional change in fault source percentage.

1.2. MANUFACTURING FAULT SOURCES

The second category of PCB faults includes those induced by the manufacturing processes of PCB assembly and soldering. Assembly faults consist of insertion errors such as components that are wrong, missing, or misoriented; leads that are bent, broken, or missing; and wires that are wrong, broken, or missing. The soldering faults are unwanted opens and shorts. Opens can result from lack of solder, improper movement of parts, or bare PC track etches of feedthrough holes. Shorts are mainly caused by solder mask and solder flow. The classic assembly fault distribution is said to be between 25 and 55 percent. This distribution can be significantly reduced by employing some degree of automatic insertion equipment with verification testing. Some manufacturers automatically insert all axial components and manually insert all other components. Others insert all ICs and manually insert all other components. Still others use automatic insertion in every possible case and manually insert the few remaining components. The trade-offs are economic. Are the costs of insertion equipment to resolve assembly faults cost effective?

Theoretically, by employing automatic insertion equipment, the fault distribution may be reduced to within the range of 10–20 percent. During wave soldering, the classic fault distribution ranges from 35 to 65 percent. This percentage may be reduced by proper care and maintenance of the soldering machine. The wave soldering machine is typically followed by 1–4 cycle washes which remove contamination and flux from the underside of the PCB. This washing is typically followed by a PCB trimmer which trims off excessive protrusions of lead ends. The purpose of trimming leads, other than packaging and safety requirements, is to ensure valid contact with a bed-of-nails test fixture. A bed-of-nails fixture is a field of spring-loaded test probes that make contact with the traces on the solder side of a PCB to obtain better electrical visibility into the PCB, (see Chapter 7). Approximately 95 percent of the multi-PCB-product vendors trim their leads, while 60 percent of the single-PCB-product vendors trim their leads.

1.3. PERFORMANCE FAULT SOURCES

The third category of PCB faults includes those resulting from weak design. Typical causes include race conditions, component interaction, excessive noise, improper outputs, and distorted signals. Performance faults may also be caused by the intermix of various vendors devices, workmanship, and variations in device specifications.

1.4. MANUFACTURABILITY

Let us consider the situation of Jack, a production engineer in a fairly large Midwestern company who is dealing with a 6802-based microprocessor controller that contains 31 digital ICs and 247 analog components. The board has an average first-pass-yield of 43 percent and a throughput of 2.3 minutes per PCB based on 250,000 units. A major ECO is introduced with the net effect of adding five more components to the PCB. Jack takes advantage of the situation and convinces engineering to design the PCB for manufacturability by changing the artwork spacing to 15 mill, and in some cases, 20 mill. He also requests enlarging the soldering pads and component insertion holes, plus orienting the components for a smoother solder flow. The result is a first-pass-yield of 87 percent with a product throughput of 1.2 minutes per PCB based upon 480,000 units. The net saving to the company is stated as being in excess of $1,500,000.

On the next-generation controller board, Jack involves himself in the engineering review in the early stages and convinces engineering to not only design the board for manufacturability but also for testability. The new controller, although more complex, contains 21 ICs and 156 analog components. The result of Jack's efforts is a first-pass-yield of 92 percent with a throughput of 0.53 minutes per PCB. The annual economical impact was estimated to be in the $10 million area. Jack, like many production engineers, believes that the best methodology to ensure high product yield and throughput is fundamentally in the design of the PCB in terms of manufacturability and testability. This suggests that every electronic manufacturer should have a production engineer on every design team or at least an active voice at PCB engineering review meetings. Does your company have production engineering involved in your new product design review meetings?

1.5. PCB FAULT SPECTRUM

Figure 1.4 illustrates the statistical PCB fault spectrum for analog, digital, and hybrid boards. For analog boards, the manufacturing process defects are low as compared with digital or hybrid boards. This fact is mainly the result of board population and the spacing of components and traces on the board, as well as the number of the device leads that require soldering. Digital boards often experience a large number of manufacturing process defects. This fact is mainly due to the high-density population of digital boards and the tightly spaced traces, plus multiple leads to be soldered for each device. As expected, the hybrid board percentages are essentially between the analog and digital board, almost proportional to the mix of analog and digital components.

With regard to component failures, note that the analog PCB has a higher failure rate than digital PCB. The majority of manufacturers

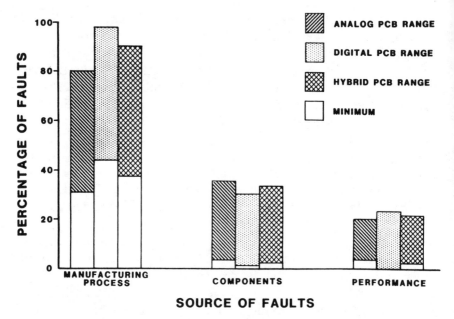

Fig. 1.4. Statistical PCB fault spectrum.

do some level of testing of digital components, but do very little testing of analog components. Essentially, more emphasis is placed on digital devices than analog devices. Why is that? The answer varies; however, there appears to be a common thread. Digital devices are complex and cost 10–20 times more than analog components. How would one justify testing a 2¢ resistor, or a 50¢ transistor with a failure rate of less than 3 percent? The world is going digital and custom hybrid devices, so why waste money on discrete analog component testing? There is an element of truth in the statement. However, maybe companies should think in terms of degrees, not all or none. Again, the hybrid board percentages are essentially between the analog and digital boards, almost proportional to the mix of analog and digital components. For performance faults, note that the analog has a significant contribution to failure. This fact is mainly due to the inherent interaction of discrete devices. The digital, on the other hand, has more failures but the minimum goes to zero when matured devices such as SSI or MSI are employed in the circuitry.

TABLE 1.2. Actual Fault Summary.

Fault Classification	Faults Per PCB				Percentage of Faults
	1	2	3	4	
Shorts	156	79	21	5	51.0
Opens	2	2	1	0	1.0
Missing Components	16	9	3	2	5.9
Wrong Components	34	27	3	3	13.1
Reversed Components	20	5	1	2	5.5
Bent Leads	31	8	3	1	8.4
Analog Specs	18	6	1	0	4.9
Digital Logic	14	8	3	2	5.3
Performance	14	8	3	1	5.1
Total No. of Faults	305	152	39	16	512
Number of Faulty PCBs	305	76	13	4	398

1.6. DEFECTIVE PCB INVESTIGATION

The purpose of the statistical spectrum is to give one an idea of the faults that can be anticipated from a particular production line. To illustrate this point, we will briefly investigate a lot of 1000 disk controller boards. The manufacturing process produces a medium first-pass-yield of 60 percent with an average of 0.5 faults per board or an average of 1.3 faults per faulty board. The actual fault summary is given in Table 1.2. The first-pass-yield is 602 boards, resulting in 398 defective boards containing a total of 512 faults.

Note that shorts represent the predominant fault. Also, the sum of percentages associated with manufacturing process defects is 84.9 percent, the sum for component defects is 10.2 percent, and the sum for performance defects is 5.1 percent.

Table 1.3 is an expansion of Table 1.2. It defines the actual faults from 1 to 4 faults per board. This example will be employed as a data base throughout this text.

1.7. FUTURE FAULT DISTRIBUTION

Let us try to predict the fault distribution of the future. It is reasonable to assume that each year production will be more successful in

TABLE 1.3. Fault Spectrum.

1 Fault Per PCB			
156 S	26 WA	21 BA	14 DL
2 O	8 WD	10 BD	6 PA
12 MA	11 RA	18 AS	8 PD
4 MD	9 RD		

2 Faults Per PCB			
31 S-S	1 S-AS	2 WA-AS	3 WA-DL
2 S-O	2 S-PD	1 RA-RD	1 BD-DL
3 S-MA	3 S-DL	1 BA-BD	1 MA-DL
3 S-WA	7 WA-WD	1 AS-AS	1 AS-PA
2 S-RA	1 WA-RA	2 MA-MD	1 BD-PD
1 S-BD	3 WA-BA	1 MA-WA	2 PD-PA

3 Faults Per PCB		
3 S-S-S	1 S-WA-RA	1 S-BA-PA
1 S-S-O	1 S-MD-BD	1 S-WD-PD
1 S-S-MA	1 S-BA-AS	1 S-DL-PD
1 S-MA-WA	1 S-DL-DL	

4 Faults Per PCB	
1 S-S-WA-BD	1 S-MA-RA-DL
1 S-WA-MA-DL	1 S-WA-RD-PD

S	Shorts	RD	Reversed Digital Part
O	Opens	BA	Bent Lead Analog
MA	Missing Analog Part	BD	Bent Lead Digital
MD	Missing Digital Part	AS	Analog Specification
WA	Wrong Analog Part	DL	Digital Logic
WD	Wrong Digital Part	PA	Performance Analog
RA	Reversed Analog Part	PD	Performance Digital

reducing in manufacturing induced faults through advances in manufacturing technology, PCB design, and innovativeness. Therefore to project, with some degree of accuracy, what can be anticipated in the future is not out of the question.

For a large manufacturing facility, Fig. 1.5 top shows a decrease in an average fault per board from 0.6 in 1982 to 0.4. in 1984, along with a decrease in the faults per defective board, from 1.4 in 1982 to 1.2 in 1984. Based on the 1000 PCB lot, this translates into a first-pass-yield of 57.1 percent in 1982, and a projected 66.7 percent in

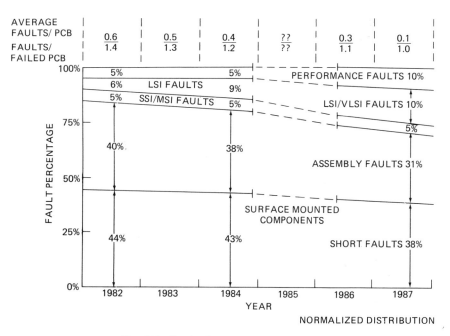

Fig. 1.5. Fault distribution projection.

1984. Further, Fig. 1.5 illustrates the shift in the fault distribution which is tabulated in Table 1.4. As one would anticipate, the relative magnitude of faults remains the same; however, both the percentages and actual faults decrease. One could anticipate by late 1984, through the first part of 1986, the common usage of two new technologies — surface mounted components and VLSI and/or custom devices.

TABLE 1.4. Fault Distribution Projections.

Fault Class	1982		1984		1987	
	%	Faults	%	Faults	%	Faults
Shorts	44	264	43	172	38	38
Assembly	40	240	38	152	31	31
LSI/VLSI	6	36	9	36	16	16
SSI/MSI	5	30	5	20	5	5
Performance	5	30	5	20	10	10
Total faults		600		400		100

Based upon manufacturing transistion from the transistor to the integrated circuit and the present manufacturers that utilize surface mounted components, the fault distribution throughout 1985 and early 1986 is totally unpredictable. However, the result in late 1986 through 1987 is predictable as illustrated in Fig. 1.5 and tabulated in Table 1.4.

In order to better understand the late 1984 through early 1986 production dilemma, let us take a few moments to discuss the transistion.

1.7.1. Surface-Mounted Components

Motivated by the success the Japanese electronics industry has experienced with surface-mounted components, U.S. manufacturers are incorporating surface-mounted components on their PCBs.

Japanese success is defined as an increase in the quality, reliability, first-pass-yield, and product throughput while reducing the PCB size and weight, manufacturing floor space, and production costs. This success translates into an increase in net profit. However, this packaging revolution requires learning five new technologies: assembly, soldering, components, PCB layout, and tester interface. Several vendors offer automatic tweezers or vacuum pick-and-place systems for tape reel, magazine, and bulk surface-mounted components. Surface-mounted components are held in soldering positions by back-to-back tape, epoxy resin, or solder paste. In the United States most surface-mounted assemblies are epoxy glass or paper laminated boards having surface-mounted components on one side and through-the-hole components on the other side. Slowly the production process is migrating to units with surface-mounted components on both sides. How best to solder surface-mounted components is an unresolved question. The most common soldering technique employed in Japan is a form of vapor-phase soldering. The printed circuit board with surface-mounted components positioned by solder paste is immersed in saturated vapors at soldering temperatures producing uniform joints, even flow seating, and minimum solder bridges. Ensuring that the components will remain in alignment when the solder paste reflows is an art.

In the United States, most PCBs contain through-the-hole components on one side of the PCB and surface-mounted components on the other side.   Component positioning is by epoxy resin for adhesive.   Soldering is by a dual wave soldering machine.   The epoxy adhesive is more popular than solder paste as it holds the surface-mounted component firm.   However, the potential of producing distorted joints, uneven seating, gas entrapment and component shadowing increases the probability of soldering defects.   Further the components are difficult to remove from the board during re-work and field repair.

Another popular soldering method is reflow soldering by baking under infrared emanating panels, producing almost the same results as vapor-phase soldering.

Surface-mounted resistors, capacitors, and diodes range from 40 to 80 mills, whereas coils, filters, and trimmers range from 140 to 240 mills.   Both have a width of 30 to 240 mills.   Small-outline transistors (SOT) range from 40 to 80 mills by 60 to 180 mills with gull-wing connections.   Small-outline integrated circuits (SOIC) are miniature flatbacks with leads at 20–50 mill centers rather than 100 mill centers.   The MSI/LSI leadless chip carriers (LCC) are flat packs with either J-bent leads or gull-wing leads.   For LSI/VLSI, 4-sided plastic flat packs (quad packs) and pin grid array packs are employed with 64 to 256 leads.   The J-bent leads and the pin grid array leads are covered by their IC package, making visual inspection and testability an issue.   PCB layout must provide the footprint for the surface-mounted components and external test pads for the J-leads and pin grid array leads.

The initial test interface to the PCBs containing surface-mounted components is a bed-of-nails clamshell fixture which makes contact with both sides of the PCB.   Another method is a robotic system consisting of a series of lift-and-place geometrically shaped probe arrays programmed to make contact with the PCB's XY-grid coordinates.

What conclusions can one glean from the packaging revolution? Standard components will remain the same.   However, the majority of electronic manufacturers will make the transistion to surface mounting in whole, or in part, within the next 2–3 years. The in-circuit tester and functional tester will evolve into the board tester with in-circuit emulation being the dominant test methodology.

1.7.2. VLSI Knowledge

Another impediment that causes a temporary distortion of the fault distribution is the amount of knowledge available about VLSI devices. Specifically, the lack of sufficient data from the manufacturers increases the likelihood that performance errors may be inadvertently designed into PCBs. This lack, coupled with the urgency of ·time-to-market, increases the probability that PCBs may be released to production before all design flaws are rectified. This tendency manifests itself in a large percentage of LSI/VLSI failures.

This description of future fault distribution gives food for thought. A recent electronics magazine article stated that the surface-mounted component market was two million chips in 1983; projected to be 80 billion chips in 1985, and growing to 400 billion chips in 1990.

2
AUTOMATIC TEST EQUIPMENT FOR PRODUCTION TEST

Production PCB testing ensures subassembly quality and monitors the manufacturing process to avoid more costly diagnostic and rework at a later stage such as system test, final test, or field repair.

Before reviewing current production PCB test methodologies, one should understand that *not* testing PCBs may be a sound economic decision. In general, such a decision will only be feasible when the PCBs are extremely simple and have a very high first-pass-yield (greater than 80 percent) or an extremely low volume (about 300 units per year). Otherwise, PCB testing is economically essential. Also remember that PCB testers will identify a component defect, but not specifically what the actual defect is. Thus, the only place the definition of a fault can be established is at the rework station after the PCB pass retest. Fundamentally, PCB test equipment is in one of two categories: either in-house or commercially available. Table 2.1 lists the various types of equipment used.

2.1. IN-HOUSE TEST SYSTEMS

2.1.1. Manual Testing

Manual testing consists of a technician using an array of instrumentation and following a test procedure to verify proper operation.

TABLE 2.1. PCB Production Test.

In-House	Commercial
Manual	Dedicated tester
Rack-and-stack	Loaded board shorts
Test station	In-circuit analyzer
Comparison test	In-circuit tester
Substitution test	Functional tester

Usually, defective PCBs are further diagnosed by a higher-level technician. Manual testing is labor intensive and involves a low capital-equipment investment that can be easily reconfigured to fit different testing requirements. Manual testing is quite common in small companies with low-volume PCB production. Relatively speaking, the test cost per PCB is very high, on the order of 6–11 dollars per fault, when you consider the unpredictable test time per unit and the variable confidence in the quality of test.

2.1.2. Rack-and-Stack

This approach uses a group of IEEE-488 bused instruments connected to a PCB test fixture and controlled by a calculator, microprocessor, or minicomputer. Typically, the test program is go/no-go, with some gross diagnostics and perhaps some aids to help the technician isolate the fault. Inherently, the rack-and-stack offers a large degree of test configuration flexibility at a moderate cost. However, the setup time is typically very long and generally requires a test engineer and a skilled technician.

The test quality is a function of the skills of the programmer. Total test time per unit is unpredictable, resulting in a fairly high per-unit test cost. Rack-and-stack testing is common in small and medium-size companies for low-volume testing and applications requiring more individual conditioning and loading for each PCB.

2.1.3. Test Station

A test station is designed and developed to perform specific tasks. The advantages include high confidence in the test quality, no excessive hardware, no unrequired function, and totally understood operation, programming, and maintenance. The disadvantages are the initially high costs of design and development and the inflexibility in reconfiguration.

The test station is fairly common in high-volume production where the PCBs require alignment or special sources and loads, or where extensive environmental testing is required.

2.1.4. Comparison Testing

Comparison testing consists of a group of sources or instrumentation that will simultaneously stimulate a known good board and the unit under test and compare their responses. This methodology has the advantage of low capital investment, reasonable setup cost, and flexibility to engineering changes. However, comparison testing is fundamentally limited to simple PCBs. As the PCB complexity increases, the confidence in the test quality decreases. Further, in comparison testing, the test and diagnostic times are unpredictable. Comparison testing is common in small companies with low production volumes of small analog or digital PCBs.

2.1.5. Substitution Testing

Substitution testing, commonly called "hot box" testing, consists of modifying the final product to allow the individual PCBs to be tested in the system. The modification ranges from simply making it easy for the unit under test to be connected and disconnected from the product, to an elaborate test fixture interface.

In general, substitute testing has a long per-unit test and diagnostic time and requires a skilled technician. The advantages include a low capital investment, high confidence in the test quality, no excessive hardware, no unrequired functions, and a thorough understanding of the operation program and maintenance. Substitution testing is popular in small and medium-size companies for both low- and medium-volume production of 10 or fewer different PCBs.

Substitute testing is also popular with third-party repair companies where the repaired PCB must be operated in the product for final acceptance.

2.2. COMMERCIAL TESTERS

2.2.1. Dedicated Testers

Dedicated testers are purchased to test specific classes of PCBs such as power supplies, memory cards, relay cards, etc. Dedicated testers have a high return on investment with a relatively fast test and diagnostic time. They are typically found where the product line involves only a few variations of one type of PCB, or is a large user of a specific item.

2.2.2. Loaded-Board Shorts Tester

When shorts are the predominant process fault, it is cost effective to employ a loaded-board shorts tester as a screener for an in-circuit functional or in-house production test system. This screening improves the test and diagnostic capability, which increases the product throughput.

The loaded-board shorts tester rapidly identifies multiple shorts in a single pass and produces fairly concise nodel failure messages to guide rework. The typical price of a loaded-board shorts tester is less than $35,000. The system software generally includes a self-learn algorithm that allows test programs to be generated and debugged in a couple of hours. Therefore, the set-up time and cost for this approach is essentially the time and cost to obtain a bed-of-nails test fixture.

2.2.3. In-Circuit Analyzer

In-circuit analyzers rapidly test for shorts, resistance, capacitance, and semiconductor junctions on an unpowered PCB. They employ a resistance crossover comparison measurement for shorts and a two-node impedance signature comparison for the other components.

An in-circuit analyzer can be used as a standalone test system or as a screener for an in-circuit, functional, or in-house production test system. This strategy represents a very cost-effective method of increasing the product fault coverage and the test and diagnostic capability, thereby increasing the finished unit yield and the product throughput.

In-circuit analyzers rapidly identify multiple faults in each of their two levels of test hierarchy. The first test level is for shorts; the second test level is for all component failures. The resulting failure message is, generally, fairly detailed. The in-circuit analyzer is typically priced at less than $45,000. Like the loaded-board shorts tester, software for the in-circuit analyzer will generally include a series of automatic learn algorithms that allow the test program to be generated and debugged from one or more known good boards in less than four days. Therefore, the setup time and cost is only marginally more than the time and cost of obtaining a bed-of-nails test fixture.

2.2.4. In-Circuit Tester

The in-circuit tester employs a guarding principle to measure the performance of individual components by electrically isolating them from the surrounding circuitry. It also detects shorts. Passive components are tested on an unpowered PCB, then power is applied and active components are tested. The in-circuit tester has a fairly large fault coverage with a specific failure message. Employing an in-circuit tester as a standalone tester or as a screener to a functional or in-house built test system is a cost-effective method of increasing the product fault coverage and the test and diagnostic capability, thereby increasing the throughput and the finished unit yield.

For the in-circuit tester, the test and diagnostic time is considerably faster than that of other testers being considered here. The in-circuit tester will identify multiple faults within each of its four levels of test hierarchy: shorts, passive components, IC orientation, and digital logic tests.

The in-circuit tester is typically priced at less than $300,000. The purchase price generally includes a software package for automatic program generation with which test programs can be generated and debugged in 2–8 weeks. Therefore, the setup time and cost are the time and cost to generate the test program plus the time and cost to obtain a bed-of-nails test fixture.

2.2.5. Functional Board Tester

The functional board tester has potentially the largest fault coverage capability because it simulates the operating environment. It produces stimuli and measures the responses of a printed circuit board in the final product environment. The functional board tester is essentially a standalone system which precedes system test. It has an extremely rapid go/no-go test time, an extremely slow diagnostic time, and it provides general failure messages which invariably require interpretation (see Chapter 3). Commercial functional testers and in-house-built substitution testers have a large degree of commonality. However, the commercial functional tester is designed for general purpose use, not for a specific range of boards, and has many software aids in both test program preparation and diagnostic execution which reduce test and diagnostic time.

Functional testers can vary in price up to $1.5 million. A typical hybrid system will cost between $400,000 and $850,000, usually including an automatic programming generator for SSI and some MSI digital devices and a simulator package for the other digital circuitry. When generating test programs for MSI and LSI digital boards, the programmer must manually insert sets of input test vectors into a simulator, then debug the resulting test program (see Chapter 3). For analog components, the programmer must manually generate the entire fault diagnostic routine. Therefore, the analog fault coverage in a functional tester is largely dependent upon the skill of the programmer.

For a functional tester, the setup time and cost must include a large allowance for test program generation, on the order of 2–6 months, plus an allowance for the UUT interface. The functional tester is a single fault detection system. Therefore, the number of rework and retest loops are a function of the number of faults per board.

2.3. LOADED-BOARD TESTING TRENDS

It is not unusual to find both in-house-built testers and commercial testers in the same production test facility. Many electronic companies migrated from in-house-built testers to commercial testers when new products are to be introduced. Others screened in-house-built testers with commercial testers. Still others have no commercial testers at all.

In 1976 Prime Data estimated 65 percent of the automatic loaded-board test systems were in-house-built. The advantages of in-house-built test systems are minimum capital investment, no excessive hardware, and no unnecessary functions. The test system's operation, programming, and maintenance are thoroughly understood by test engineers who were utilized to accomplish the task. The disadvantage was the time and cost of developing and manufacturing the test systems, lack of documentation, and the inflexibility of the test system. As the commercial ATE vendors broadened their test systems capabilities, offering a modular general-purpose tester with user-friendly software, companies started requiring build-or-buy justification. Needless to say, when the cost of engineering and manufacturing setup is amortized over a large number of testers, the financial and delivery advantages become evident. The in-house-built, loaded-board test systems were estimated at 50 percent of the systems in use in 1981, 43 percent in 1983, and projected to be 33 percent in 1985. The percentage of users of in-house-built test systems versus commercially purchased testers varies with the industry. The instrumentation industry is the largest user of in-house-built test systems at 57 percent, projected to be 67 percent in 1985. The aerospace industry is the smallest at 30 percent in 1983, projected to be 25 percent in 1985.

The loaded-board test system at present requires 1500 points or less for in-circuit testing, 5000 points or less for continuity testing, and 400 points or less for functional testing. Again both the computer and computer peripheral industries are the exception, requiring 25 to 30 percent more test points, while consumer and medical industries, as a maximum, require 25 to 35 percent fewer test points. The present digital test rate is 250 kHz to 5 MHz for in-circuit testing and 1–10 MHz for functional testing. Another trend stimulated by surface-mounted components is the migration from service repair to production's functional testing employing in-circuit microprocessor emulation as the source with both signature analysis and standard functional vector testing pin electronics as the sensor. Two main advantages are in test program preparation and real-time testing. Functional emulation is discussed in Chapter 8. The 1985 projection is that digital in-circuit testers will require a 1–10 MHz test rate and functional testers will require a 10–30 MHz test rate.

For test fixturing the most popular sizes of PCBs are 8 by 10 inches and 5 by 5 inches. The computer and computer peripheral industries are the exception, with 15 by 15 inch and 14 by 18 inch PCBs. At present 60 percent of the industry employs a standard 100-mill grid. The 1985 projection is that 84 percent of the companies will employ 25, 50, and/or 100 mill standard PCB grids. This suggests the test system will require CAD/CAM interfaces as a test program generation aid.

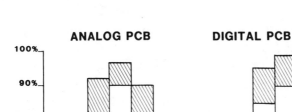

Fig. 2.1. Typical fault coverage.

2.4. TESTER'S FAULT COVERAGE

In a given production test facility, it is not unusual to see various combinations of both in-house and commercial test systems for implementing the company's test strategy. Fig. 2.1 illustrates the typical fault coverage for shorts testers, in-circuit analyzers, in-circuit testers, and functional testers for analog, digital, and hybrid PCBs. As shown, the in-circuit tester will consistently provide the highest fault coverage for analog PCBs, while the functional tester has the dominance in fault coverage capability for digital PCBs. As expected, the in-circuit tester shows dominance for hybrid boards. Needless to say, with effort, functional analog testing can be better than in-circuit analog testing. However, few companies found the effort to be cost effective.

Viewing fault coverage capability from a financial aspect, the in-circuit analyzer has a significant cost advantage over the other test equipment being considered. This statement is also valid for digital

PCBs, but the in-circuit analyzer tests are for unpowered PCBs and no digital logic verification is performed. When a static test of a digital PCB is adequate to ensure its functionality, an in-circuit tester has a significant cost advantage over a functional test system. If the digital PCB requires high-speed or dynamic testing to ensure its operation, only the functional test systems can accommodate the test requirements.

The typical fault spectrum is a function of the type of PCB to be tested. The total PCB fault spectrum minus the tester fault coverage equals undetected faults. If the test system has 98 percent fault coverage, and the uncovered 2 percent are the cause of the PCB failures, all the PCBs will fail in system test. In-circuit analyzers, in-circuit testers, and functional board testers have large individual variances between their test specifications. Several vendors offer a series of similar systems, with each test system designed to satisfy the requirements of a specific market. Further, in studying a PCB schematic, one will often find components that are masked from fault detection by one type of test methodology but detectable by another. Or the faults may be undetectable at the PCB test level. The fault coverage of a given test system will always be a function of the fault spectrum of the PCB to be tested.

Estimated fault coverages are confirmed by the results of fault detection at the next test station, assuming the next test station is capable of detecting the fault. Each test system algorithm has a series of tests that are performed to indicate each fault class. For example, one test system will perform three tests on a transistor while another will perform five. In each case, if the transistor test passes, it is defined as good. The thoroughness with which the tests are performed reflects the quality of the test algorithm. The completeness of the test program is referred to as the quality assurance (QA) rating and is a measurement of the ability of the test program to test all possible fault classes on the UUT as provided in the test algorithm. Completeness plus thoroughness equals the test system's effectiveness. The higher a test system's effectiveness, the higher the next test station yield.

2.5. FAULT COVERAGE AND TEST PROGRAMMING

Figure 2.2 illustrates a comparison of the typical ranges of fault coverage and test programming time for each of the four prime PCB

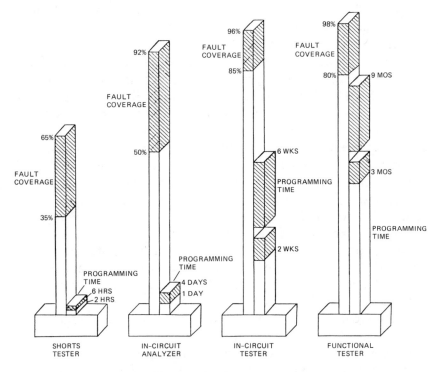

Fig. 2.2. Fault coverage.

test systems. In each case, the fault coverage is a function of the type and complexity of the PCB. Typical fault coverage percentages for digital, analog, and hybrid PCBs are illustrated in Fig. 2.1. Setup time and money are critical criteria when considering the selection of a test system for a new product, as is the flexibility of the test system to test different PCBs as the manufacturing load shifts. Generally, time is more critical than money.

As illustrated in Fig. 2.2, the shorts tester has a very rapid programming time – a matter of hours, primarily because of the self-learn capability from a known-good board. The in-circuit analyzer provides greater fault coverage but the typical programming test time is 1–4 days. The in-circuit tester provides an even greater fault coverage, but its typical programming time is 2–6 weeks. It should also be noted that, if a complex IC library subroutine is required and is not included in the in-circuit library, the programming time may be extended by an additional 6–8 weeks to develop the subroutine for that specific device.

The functional tester has a typical fault coverage of 80–98 percent, requiring a typical programming time of 3–9 months. Again, this time may also be extended if the functional IC library does not contain a required device model. The IC library element in a functional board tester is different from the IC library element in an in-circuit tester. The functional board tester's IC library element is a mathematical model of the IC device. The in-circuit tester's IC library element is a subroutine consisting of a series of tests that cause all the inputs and outputs to be tested for stuck-at faults.

2.6. TEST DIAGNOSTICS

Figure 2.3 illustrates typical test times of the four prime PCB testers under four different test conditions. For go/no-go testing, the shorts

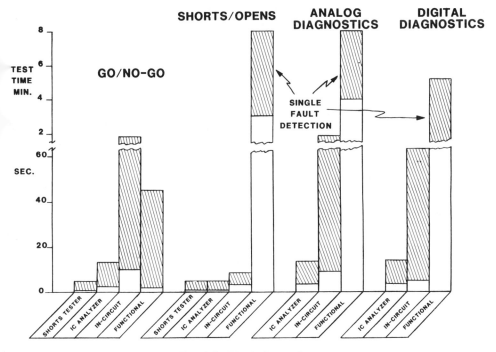

Fig. 2.3. Typical test time.

testers, in-circuit analyzers, and functional testers are extremely fast. The in-circuit tester is slower because the go/no-go test time is the sum of the diagnostic test times at each level of its testing hierarchy. A loaded-board shorts tester has one test level: shorts. The in-circuit analyzer has two levels: shorts and unpowered components. The in-circuit tester has three major test levels: shorts, unpowered components, and powered components. This last level is further subdivided into either two or three levels: IC orientation, digital logic, and in some cases, analog specifications.

For in-circuit testing, the diagnostic time is extremely short compared to functional testing. Long diagnostic times with functional board testing are due to the operator intervention and probing which is generally required to localize a fault or a number of possible faults (called a fault net). The mean average is 12–32 probes for a short, 15–32 probes for an analog component, and 8–20 probes for digital logic. Further, the diagnostic capability of a shorts tester and an in-circuit analyzer or an in-circuit tester is multiple fault detection within each level of its test hierarchy, whereas a functional tester has a single fault detection capability. The multiple fault detection capability provides faster fault diagnostics.

If a PCB contained three independent shorts, one test by an in-circuit tester would detect all three shorts. The functional tester would identify one short; after that was cleared it would identify the second short; and after that was cleared it would identify the third short. Therefore, when a functional board tester is used, maximum throughput is obtained when the board contains a minimum number of faults. This is the goal of a high-volume production line.

Another factor affecting throughput is PCB handling time. It is not unusual for a shorts tester, an in-circuit analyzer, or a functional tester to have a test time shorter than 10 seconds, which is typically less than the PCB load-and-unload time. A dual chamber fixture provides two test fixtures under separate vacuum control. This capability allows one PCB to be loaded on one test fixture while the PCB already on the other test fixture is being tested. By the time the test is completed, the newly loaded PCB is ready for test. This ping-pong action effectively eliminates handling time. For an in-circuit tester, the total test time is generally greater than the PCB handling time. Therefore, by employing a dual chamber fixture, the effective handling time is zero and the total PCB throughput on an in-circuit tester could be equal to the total PCB throughput on a shorts tester, in-circuit analyzer, or functional tester.

For functional testers, diagnostic time is in excess of most acceptable limits. Therefore, to minimize functional board tester's diagnostic time, PCBs should be screened by an in-circuit tester or analyzer before being tested on a functional board tester, which can then concentrate on performance faults in a high first-pass-yield situation. The only methodology which appears practical to improve throughput on a shorts tester or in-circuit analyzer is some form of rapid automatic board handling system.

2.7. UNVERIFIABLE FAULTS

Rob is a test engineering manager with a medium-sized West Coast company. One of his recent assignments was to select an in-circuit tester for their production line. He narrowed the field to three vendors and proceeded to do a detailed matrix evaluation based upon vendor presentations and system demonstrations. Unable to make a comfortable decision, Rob elected to run a test case, commonly called a benchmark, between the three vendors. The result was that Rob selected Vendor A because his in-circuit tester caught more faults than the other two vendors. Vendor A caught 363 faults, Vendor B caught 344 faults, and Vendor C caught 351 faults. Vendor A was also 45 percent faster than the others. Vendor C objected strongly to the results and requested Rob, for his own sake, to verify the 363 faults that Vendor A detected. Rob agreed, and, under controlled conditions and employing his present test strategy, conducted a test of the 75 PCB evaluation lot. The results were: total failures amounted to 352, of which, Vendor A detected 346, Vendor B detected 341, and Vendor C detected 349. That is, the unverifiable faults were: 17 for Vendor A, 3 for Vendor B, and 2 for Vendor C.

Vendor A's test methodology was not suited for a significant portion of Rob's fault classes. Vendor B exhibited very good quality of test and Vendor C's quality was even a little bit better. Rob could rationalize why the various test systems missed specific faults — tolerance too loose, not enough wait time, logic levels set too loose, etc. However, initially Rob had a problem with accepting a tester that failed the wrong part or, for no apparent reason, failed a part that was in no way associated with the actual failure.

Rob conducted an investigation into the probable causes of unverifiable, or bogus, faults. The first area of probability was improper or inappropriate unit-under-test (UUT) tester interface. Possible causes of bogus faults on the UUT and in the test fixture system could be:

- PCB contaminated solder pad(s)
- Excessive component lead(s)
- Dirty or contaminated test probe head(s)
- Inappropriate test probe head type(s)
- Bent or misaligned test probe(s)
- Weak or broken test probe spring(s)
- Poor mechanical registration
- Insufficient test fixture actuation
- Noise and crosstalk
- Inductive and capactive effects

There is no way to guarantee that the above causes will never produce an unverifiable fault. However, several simple steps can be taken to minimize the probability.

The PCBs nodes must be free of solder flux, contaminants, and other debris to ensure reliable and repeatable test probe electrical contact. Commercial circuit board washes have proven quite reliable. Some low volume manufacturing facilities have employed heavy-duty dish washers with great success. Further, lead trimming is an obvious corrective action to eliminate excessive lead length. A router does an excellent job.

The most predominant cause of unverifiable faults is contaminated test probe heads. The test fixture's probes are electrically tested by employing a test probe verifier and performing a resistance test. The test probe verifier is unetched bare board except for the ground and power supply traces. The ground and power supply traces are connected to the unetched trace plane by small resistors, typically 200 ohms, forming a resistance driver. Each test probe is individually selected in measuring each resistor. Any variance in the measured and stored resistance value indicates a possible buildup of contaminants or dirt on the test probe head or a catastrophic failure.

Test probes with gold-plated plunger and head have repeatedly shown the best performance in terms of consistently low contact resistance throughout many thousands of operations.

Test probe mechanical compression should be tested periodically. A strain gage should read 8 ounces at 75 percent of the probe travel. Note that on occasion, 4-ounce test probes are employed for 50- or 10-mil center artwork. A simple test is to depress the test probe in question with a known good test probe. If both test probes compress equally, the test probe in question is good. If the test probe in question compresses more than the known good test probe, the spring is weak and the test probe should be replaced.

The bent or misaligned test probe generally manifests itself as a random failure. With each repeated test fixture actuation, the bent probe rotates in a random direction and periodically misses the target node pad. The misaligned test probe randomly slides off the target node pad. The inappropriate test probe bounces off or does not cut the oxide film of the target node pad, making poor electrical contact.

Catastrophic test probe faults are detectable by employing a UUT/Test Fixture Verification software routine before each test, or when a failure is detected, to ensure proper test probe electrical contact.

To test the UUT, every PCB test node must accurately align with the respective test fixture's probe. Tooling holes in the PCB and matching tooling pins in the test fixture help to guarantee this registration, assuming the test probe is properly installed in a true vertical and is accurately positioned relative to the tooling pins. Once a UUT/Tester Interface has been proved, the registration problem appears to be most susceptible to wearing tooling pins and/or enlarged or miscentered PCB tooling holes.

Insufficient test fixture actuation may not provide enough force for the test probe to cut through the target node pad oxidization, thus making poor electrical contact, or the vacuum CFM may not be sufficient to hold the PCB firmly on the test probes during the test sequence. Again, UUT/Test Verification may help identify this problem when it occurs.

A major safeguard is to store the test fixture in a clean environment with a protective dust cover mounted on the test probe field.

The test fixture should have its own preventive maintenance schedule. After a predefined number of test operations or a defined time interval, the test fixture's probe heads and gasket should be cleaned by wiping and/or brushing with a solvent solution.

The test fixture should be visually inspected and electrically tested at the beginning of each shift. The inspection should include the test probes, gasketing, vacuum hose, clamps, and port, and an accelerated thrust followed by a smooth settling, actuation, and hold operation.

Noise, crosstalk, capacitive and inductive effects are possible causes of bogus faults when performing digital tests. The test fixture can easily contain more than 500 wires of lengths up to two feet. This, plus the tester, will load each PCB node with 75 to 250 picofarads of capacitance. This loading, plus inductive coupling, is a probable cause for a bogus fault. Whenever high speed, parallel burst stimuli is employed, the possibility of excessive coupling within the test fixture wiring exists during pulse transition. As a minimum, twisted-pair should be employed on each node. Connect the ground of the twisted-pair to the test-fixture ground plane. Assuming a ground plane is present, ground the other end of the twisted-pair ground wire to the tester ground. Do not neatly bundle or cable wrap the twisted-pair wires. Looking neat could promote crosstalk.

Another area of probable unverifiable faults is in analog measurement. The possible causes could be:

- A failed component causing other good component failures
- Inadequate tolerance allowance
- Erratic or unstable measurement
- Insufficient or improper guarding
- Inherent component variables
- Improper measurement

An in-circuit tester is not a component tester. Therefore, test engineers should not have a mental block in exploring causes and solutions outside of the specified component's value and tolerances.

In the majority of cases, when a failed component causes good components to be identified as failures, the real component failure is a semiconductor. The automatic program generator regards semiconductor, and in some incidences capicitors, as current disabling devices. When a resistor is guarded out of a circuit, virtual ground potential is present at both terminals.   If the resistor shorts, other component measurements will not be affected.

However, in some cases, when a semiconductor shorts, an additional current path is created.   This could distort other component's measurements associated with the semiconductor's circuitry.   The result would be failure string message like Q13-shorted B/E, R14-HI, C21-LO or CR-17-shorted.   As a general rule, when a test results in a failure string or multi-failures contained within the same circuitry, the probability is that the semiconductor is the only real failure.   If a semiconductor is not present, a capacitor is next in the order of probability.

To minimize this type of bogus fault, exploratory work has been conducted in the area of active guarding, that is, guards at a voltage potential other than common/ground.   Preliminary work also has been conducted regarding the use of incorporating "what if conditions" in the APG so that when a failure occurs, a branch to a diagnostic algorithm is implemented.   However, to date such efforts have been manually performed or attempted by the test engineer with the aide of test program verification software routines.

An example of inadequate tolerance allowance is:

A series of PCBs, say 50, have been tested and a high failure rate occurs on a specific component.   First, investigate the stability of the measurement by repeating the testing of the same PCB 10 to 15 times while monitoring the measured value and variances.   Assume, for this example, that measured values are stable (within a couple of percent) and are clustered just above the upper tolerance limit.   The component in question, then, is removed and measured out-of-circuit.   The component value was just within the upper component specification limit.

Investigating the same component on a PCB that previously passed showed the component measured value after repeated testing was stable, again within a couple percent.   The measured values were clustered in the middle of the tolerance range.   This component then was removed and measured out-of-circuit.   The component value was just above the lower component specification limit.

In both cases the components are good.   However, when measured in-circuit, one passes and one fails.   The conclusion is the component test tolerance limits did not take into account the tester's measurement tolerances and effects of test fixture wiring with relationship to the component specifications. The first recommendation is to change the nominal component value in the amount of the average offset. A second solution is to expand the upper test tolerances limit while reducing the lower test tolerance limit.

Take the same case, except the component failed in-circuit just below the lower tolerance limit. The component out-of-circuit measured value was in the middle of the specification range. The same component on a previously passed PCB measured in-circuit values clustered just above the lower tolerance limit, but upon removing and measuring out-of-circuit, the component value was just below the upper specification limit. Again, in both cases the components are good. When tested in-circuit both measured low and one passed and one failed. This indicates a negative in-circuit measurement offset that may be caused by a small shunt fixed current, fixed guard loop current, or tester tolerance. The first recommendation is to change the nominal value in the amount of the offset. The second solution is to expand the lower test limit while reducing the upper test limit.

Assume the case of a repeatably stable in-circuit measured component that tests good out-of-circuit with measured values on different PCBs of below, through, and above the test tolerance limits. The solution is to expand both test tolerance limits.

The erratic or unstable measurement occurs when a component on a PCB repeated measurement has a large variation in values. This could be the result of improper measurement delay, or insufficient or improper guarding. The measurement delay is employed to ensure that a steady state condition is obtained before a measurement is made. For example, measuring a capacitor before the capacitor is fully charged. The measurement is taken on the rising slope of the voltage response curve. When a measurement is made before the voltage response is stable, an erroneous value is obtained. Further, the value is rarely repeatable. The obvious solution is to increase the measurement delay. If sufficient measurement delay cannot be obtained, the component can be tested only by calculation. The measurement delay also is employed to allow time for a capacitor to discharge before a stable measurement can be obtained.

The effect of insufficient or improper guarding is the availability of undesirable shunt current paths that result in erratic measurements.

For insufficient guarding, a possible solution — before making the decision that the component is untestable — is to multi-guard component nodes whose opposite terminals are connected to the desired guarded node. This is an attempt to increase the guard ratio.

Some components have unique properties that vary with the environment, temperature, humidity, or current. A carbon composition resistor is a good example. When the carbon resistor absorbs moisture, the resistance value increases. When the carbon resistor dries out, the resistance drifts back to normal.

Unique property components can be very frustrating to a test engineer because of their rare usage and component properties are the last points investigated.

Improper measurement is simple but sometimes hard to find, comparable to troubleshooting a system with a blown fuse, measuring a resistor as a capacitor, or a diode as a transistor.

Not all components can be isolated from their surrounding circuitry for measurement, like components in parallel can not be measured individually. Capacitors in parallel will be summed. Depending on their number, value, and tolerance, the in-circuit tester may or may not be able to determine if one is missing. The in-circuit tester will detect a short but is unable to identify which capacitor. For a diode in parallel with a resistor, the tester can determine the diode's orientation, if one component is missing or shorted. Other limitations for component isolation and measurement are generally a function of the in-circuit measurement capability and guard ratio.

The third area of probable bogus faults is in digital testing. The possible cause(s) could be:

- Oscillations or transmission-line effects
- Insufficient or inappropriate backdriving
- Variance in IC brands
- Improper sense strobe position
- A failed IC causing other good IC failures
- Improper IC output isolation
- Inappropriate UUT power
- Inappropriate library element

An in-circuit tester is not a digital component tester. The goal of the in-circuit tester is to ensure the correct IC has been installed on the PCB, that it is free of pins faults and basically functions correctly.

As a general rule, to avoid serious oscillation, totally disable any on-board oscillators. Disabling the output oscillator gate often is not enough to stop leakage and crosstalk. The preferred practice is to ground the oscillator's output with a relay mounted inside the test fixture to a high-quality ground.

Digital component isolating consists of overdriving the UUT inputs to obtain total input stimulus control and inhibiting any common IC outputs from interfering with monitoring the UUT output responses. This is accomplished by overdriving the other device's inputs to provide the desired output state. Overdriving a device input is technically backdriving the previous logic stage's outputs.

Insufficient backdriving is where the current capacity is inadequate to force a change to the proper logic level on the output of the device. Take the example of a bus driver as an input to the UUT, with each bus driver requiring a 450 Ma current backdriver to force the proper logic level and the system's capability is 400 Ma. In this case the bus driver inputs should be overdriven to provide the desired logic output state. This provdes assistance in backdriving the bus drive, and controlling the inputs stimulus of the UUT. This is referred to as multi-guard.

Inappropriate backdriving occurs when the backdriving current pulse causes a glitch on the line which triggers the UUT. An example is directly trying to disable a feedback loop or a clock line. The appropriate method is to disable an intermediate gate that inhibits the control line in question. If an intermediate gate is not present, UUT isolation can be a serious problem and design for testability would become an issue. Testability is the ability of isolation, control, measurement/sensing, and accessing a circuit for testing.

In testing CMOS devices, if the logic levels are too high (approaching or beyond VDD) the CMOS device latches. The crowbar effect will cause the device to rupture or blow up.

Improper sense strobe positioning is programming the sense strobe too close to either end of an event, resulting in an erroneous value. The cause of the erroneous reading could be that the stimulus did not have time to propagate to the output before the sample was taken or it was caused by sampling noise because of the logic transition and transmission effects. The solution is to delay the sense strobe so that the event is approximately 75 percent completed before the sample is taken.

When testing digital ICs in-circuit, the overdriven UUT inputs will mask most input faults, except catastrophic internal faults or a dead short. If an IC has an input fault, there is a good chance it will not be detected when testing. However, when testing the previous logic stage, the output will be loaded by an input fault in the next stage. Therefore, the driving logic stage output will fail proper logic level. On replacing this IC, the problem will remain as if you just replaced a good IC device.

A possible solution is to confirm output level failures before replacing the device.

Inappropriate UUT power is generally defined as not enough current capability to power the PCB. This loads down the power supply, dropping the voltage, which in turn causes the voltage-sensitive devices to fail. Other possibilities include the total PCB not being under power or the wrong voltage being applied.

The IC library element is a test subroutine to test a specific IC device identified by number and brand. Equivalent IC devices will not necessarily have the same test quality as the IC used to verify the test subroutine. Further, some equivalent ICs have additional functional and/or different I/O architecture. The result could be a bogus fault.

As the UUT's inputs are forced, the UUT's output response is being sensed and compared to storage threshold values and logic state. To correctly monitor the UUT's outputs, any other ICs outputs must be in a noninfluential state. If any IC is improperly disabled, the UUT may be erroneous and the UUT will fail.

The last area of probable bogus faults is the computer section and operator:

- Brown out or power transients
- Poor software design
- Poorly maintained storage media
- Operator not following VDT prompts
- Improper seating of the PCB

Low line voltage or power transients may cause the memory to drop bits, change bits, or the system may hang up completely. Poor software design refers to software not totally debugged or software with many unqualified branches.

In the case of a floppy disk transfer of main storage media, care should be taken not to write on the diskette with a lead pencil. Further, the diskette will wear and drop bits and/or jump to different sectors.

The test system communicates to the operator by indicator lights and video display. The system requests the operator to perform an action or probe specific PCB nodes. When completed, press continues. If the operator performs this incorrectly, the tester measures the result and either fails the test or travels along the wrong diagnostic algorithm. When the operator places the PCB on the test fixture, there is a tendency to force the PCB or part of the PCB in a direction in the name of assistance. This may result in poor registration and component failures.

Rob's conclusion was that unverifiable faults are present in any test system due to a system glitch, a fixture glitch, or an operator glitch. Thus, it is necessary to determine the acceptable level of unverifiable faults.

Rob conducted a survey of his peers and, to his surprise, found a large number did not realize that their in-circuit testers were failing good parts until the issue was presented by Rob and verified by his peers. The range of 0–2 percent unverifiable faults was agreed upon as an acceptable level of pain. This by no means is an acceptable standard, merely a meeting of the minds of a group of people. However, the point is that unverifiable faults exists.

From the in-circuit benchmark, Rob learned on analyzing his data base that 15–25 PCBs containing known faults are quite sufficient for any benchmark. All in-circuit testers are not the same. A test methodology may be acceptable for one type of PCB and totally unacceptable for another type of PCB, and there is a production integration time for any tester where its value thresholds are tuned to meet the need.

2.8. FAILURES AT SYSTEM TEST

A common question is, after implementing an effective PCB test strategy, what can we anticipate as a failure rate at system test?

TABLE 2.2. System Test Failures (in percent).

Industry	Average Failures	Failure PCB Mix		
		Digital	Analog	Hybrid
Computer	3.3	47	28	25
Peripheral	3.6	31	30	39
Office/Business	2.6	33	35	32
Instrumentation	3.8	30	39	31
Communication	3.5	37	36	27
Aerospace	3.3	35	33	32
Military	3.4	29	35	36
Consumer	2.4	30	47	23
Industrial	3.1	34	36	30
Medical	3.8	33	39	28

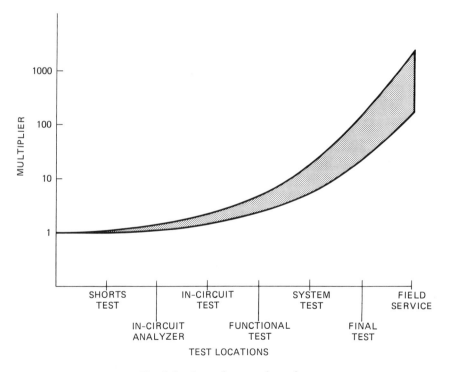

Fig. 2.4. Cost of test and repair.

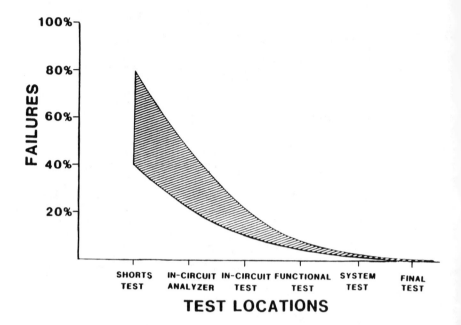

Fig. 2.5. Product failures.

Because of the large number of variables a general answer is not available. However, in late 1983, a Fairchild survey was conducted to ascertain what the various industry segments which employed a PCB test strategy were experiencing at system test. Table 2.2 lists the results. In the computer industry, the average failure rate at system test was 3.3 percent, of which 47 percent of the PCBs were digital, 28 percent were analog, and 25 percent were hybrid. In scanning the survey data base, the system test failures for digital PCBs varied from a high of 4.7 percent to a low of 2.2 percent, 4.4 to 2.7 percent for analog PCBs and 3.4 to 1.7 percent for hybrid PCBs. One may consider these results as a typical range of anticipation.

2.9. COST OF TESTING

Starting with the board test facility, Fig. 2.4 illustrates the relative cost of testing and repairing faulty PCBs for different types of testing locations. Note the exponential increase in cost. Logic would say the majority of faulty PCBs should be repaired at the minimum cost. Figure 2.5 shows the reduction in faults before they go to a more expensive and slower tester. This strategy means the more complex testers are not burdened with many faulty boards so they can handle a large volume of boards and can concentrate on diagnosing the problems of complex devices and design irregularities. Chapter 4 will discuss various test strategies with regard to average test time, throughput, and cost.

3
PRODUCTION PCB TEST
SYSTEM COMPARISON

For over a decade prior to 1978, the functional board test system, whether in-house built or commercial, was the predominant methodology for production printed circuit board testing. Then conditions began to change as two events occurred. First, the mechanical and electromechanical industries recognized that, with the proven capability of the microprocessor and other LSI technology, their products could be more controllable, repeatable, flexible, and reliable with significantly lower manufacturing costs. Further, the electronic manufacturers were concurrently designing and building more complex boards.

Both groups faced the same production testing problems: high cost, lengthy time, and high technical skill level required to generate the PCB test programs and lacking library models, compounded by the high cost, lengthy time, and high technical skill level required to perform the testing, diagnosis, and rework tasks. The search for a more cost-effective test strategy resulted in an upsurge of the in-circuit tester market. It should be clearly understood that an in-circuit tester does not replace a functional board tester. Each type of tester has its advantages, and the most effective test strategy for manufacturers involves combining the strengths of both in-circuit and functional board testing.

3.1. IN-CIRCUIT AND FUNCTIONAL BOARD TESTER COMPARISON

Table 3.1 shows a condensed comparison between an in-circuit tester and a functional board tester. The in-circuit tester is a manufacturing verification tool. It tests individual components and interconnections on a printed circuit board. It is assumed that, if the board design is proven and the artwork and components are correct, the board will be operational. The validity of this assumption is a

TABLE 3.1. In-Circuit/Functional Test System Comparison.

Topic	In-Circuit Tester	Functional Tester
Type of testing	manufacturing verification	performance
Test methodology	individual component	stimulus/response
Go/no-go test speed	slow	fast
Diagnostic time	fast	slow
Type of fault detection	manufacturing faults	engineering and manufacturing faults
Fault detection criteria		
Analog components	device specification	output signal
Digital components	pin faults	logic results
Fault isolation	component level	nodal
Fault resolution	component/component net	nodal/nodal net
Fault accuracy	high	medium
Fault detection capability	multiple	single
Logic timing detection	device level	total board
Typical fault coverage		
Analog PCB	90–95%	65–90%
Digital PCB	85–94%	90–97%
Hybrid PCB	87–94%	75–92%
Unverifiable faults	low	very low
Failure message	specific	general
Technical levels required		
Test operator	low	low
Diagnostics operator	low	medium
Programmer	medium	high
Maintenance	medium	medium
Test programming effort		
Analog circuitry	low (1×)	high (20×)
Digital logic	low (1×)	high (8×)
Debug	low	high
Integration	medium	low
Learning curve	3–6 mos.	6–12 mos.
Programming aides	APG	logic simulator
100% utilization	4–9 mos.	9–18 mos.
ECO difficulty	low	high
Capital cost	medium	high
Programming cost	low	high
Fixture cost	medium	low
Ownership cost	low	high
Reliability	high	high
MTBF	medium	medium
MTTR	low	medium

function of the electrical designer's safety margins and the PCB complexity, i.e., using 100K ECL devices. During test, the functional board tester duplicates final product environment by stimulating the board inputs and measuring the outputs, thereby testing board performance.

Testing the functionality of a good board is faster than performing in-circuit testing of the individual components on a board. Basically, functional go/no-go testing for a board with 175 digital ICs can be performed about 6–8 times faster than in-circuit testing. Therefore, a functional test system is an ideal tool for high-yield/high-volume production. That is, if an in-circuit tester were employed as an in-line test system for this application, there would be a significant reduction in throughput.

The in-circuit tester detects faults caused by the manufacturing process and the majority of component faults. The functional board tester detects faults caused by engineering, manufacturing, and most component problems. Therefore, for performance fault detection, timing or interactive component failures, the only method of fault diagnostics is functional board testing. In the majority of cases, the component-under-test can be electrically isolated, the in-circuit testers define a fault to the component level. When a component cannot be electrically isolated, the in-circuit tester either defines a no-test or a component net failure. Further, a component beyond the in-circuit tester's specifications is commonly defined as a pass or fail, which generally results in the test being aborted. With the exception of simplistic circuitry, the functional board testers isolate a fault to the node level. When multiple nodes are defined as possible faults, the functional board testers go into a computer-guided probe mode. The operator is directed to probe numerous points on the board in an attempt to isolate the multiple node fault to a single node fault. After probing an average of 9–38 points, the result may be a list of probable faults. This operator intervention is quite time consuming and introduces the possibility of human error.

The in-circuit tester on the other hand, will detect multiple faults simultaneously within each level of its testing hierarchary (see Chapter 6). That is, if there are four independent shorts on a board, one test will detect all four shorts. The functional board tester performs a single-fault detection. Thus, if there are four shorts on a board, the board would have to be tested four times to find all four shorts. It is not unusual for a board with four shorts to take about 5 seconds test time on an in-circuit tester and as much as 10–15 minutes test time on a functional board tester. An in-circuit tester has no way of determining any critical timing defects such as race conditions within a logic path. The functional board tester may detect these failures. The in-circuit tester verifies individual components and is not concerned with the interaction of devices, whereas a functional board tester verifies overall board performance by promulgating test patterns through all the paths of interest.

The diagnostic test time for an in-circuit tester is fast, always less than the good board total test time, except when the fault is contained in the last level of its test hierarchy, which results in equal test time. The diagnostic test time for a functional board tester is extremely slow, generally several orders of magnitude greater than its good board total test time. A functional board tester identifies one fault at a time, and to a lesser degree, commonly requires human intervention in probing.

For analog testing, the in-circuit tester isolates each component electrically and then measures the component with respect to a stored value and tolerance derived from the component specification. The functional board tester stimulates the analog inputs to the board and measures the resulting signal outputs. A component may be marginally out of specification and not affect the PCB performance, thereby passing the functional test. However, the in-circuit tester as it tests individual devices to specification will identify the component as a failure. Note, this may mean the in-circuit tolerances are set too tight and should be relaxed. For analog functional testing, instrumentation is added to either a digital functional tester or an in-circuit tester. The functional board tester has a larger capability in this arena. In-circuit testers have the limiting factor of the switching matrix plus fixture capacitance. Therefore, extensive analog functional testing is discouraged in the in-circuit world and encouraged in the functional world.

For digital testing, the in-circuit tester employs backdriving techniques to control the digital device inputs and digital guarding techniques to isolate the digital device outputs, then stimulates the inputs and senses the outputs, looking for pin faults (see Chapter 6). The functional board tester, in contrast, stimulates the digital inputs with a series of test patterns and senses the board output for logic failures. Most SSI and low levels of MSI device logic test patterns are identical in component, functional, and in-circuit testing. However, in testing an LSI device with a component tester, it is not unusual for the test program to contain in excess of 40,000 vector sets. For a functional board tester, the same device library model typically would have less than 2,000 vector sets, and for an in-circuit tester, the device test program would typically have less than 500 vector sets. In component testing, all the internal logic functions of a digital device must be qualified at various voltage levels and the input/output pins must also be verified for voltage and current capability. The same library model in a functional board tester is built to the logic specification on the vendor's data sheet. After initialization, each input/output pin will make a one-zero-one transition as prescribed by the device logic equation. The same library subroutine in an in-circuit tester is programmed to verify the one-zero-one transition capability of the device input/output pins by the most expedient method. The number of test vector patterns is directly proportional to the extent the device is exercised.

The fault coverage of an in-circuit tester is a function of the unit-under-test (UUT) fault spectrum and the ability of the in-circuit tester to isolate the specific components electrically. The fault coverage in functional testing is a function of the complexity of the board, the UUT fault spectrum, and the number of components that affect board performance. For a digital board, typical fault coverage (shorts and stuck-at-faults) of an in-circuit tester is 85–94 percent; whereas functional testing the same board would result in a fault coverage of 90–97 percent. This fault coverage indicates that functional testing has a much higher level of effectiveness in testing a digital board.

However, for a pure analog board, an in-circuit tester has a typical fault coverage of 90–95 percent; while functional testing of the same board would result in a fault coverage of 80–90 percent. An important question is, how does one determine the analog fault coverage of the functional board tester test program, as there are no fault models or trackable conditions where a quality assurance rating can be obtained? This suggests that the only methods available are manually inserting known faults and identifying same or incorporating tests to detect failures that were detected in the next test station. The functional board tester yields a lower fault coverage primarily because the functional board tester is only concerned with the cause-and-effect relationship of inputs and outputs, and the test program is only as good as the skill of the programmer.

In-circuit testing of a hybrid board typically gives a fault coverage of 87–94 percent, while functional board testing gives a typical coverage of 83–95 percent. These percentages will vary greatly as a function of the mix of analog and digital devices on the board. The more digital components, the higher the effectiveness of functional testing. The higher the number of analog components, the greater the effectiveness of in-circuit testing. It should be noted that there is a large overlap in the fault spectra where faults are detected by both in-circuit and functional testing systems.

The failure messages issued by in-circuit testers will typically specify the particular component that caused the fault or the specific failure mechanism, whereas failure messages from a functional board tester declare nodes or multiple failure probabilities. Specific failure messages significantly reduce the time to localize the actual fault at the rework station. A detailed discussion on the causes and effects is contained in Chapter 5.

The functional and in-circuit testers both require essentially the same technical level for the test operator. However, the functional board tester's operator requires more training, as the operator must be able to follow computer instructions for probing.

The major area of difference between in-circuit and functional board test systems is in the generation of test programs. In-circuit test systems use an automatic test generator whose input data base is simply the UUT part list and interdevice wiring. Therefore, only a rudimentary knowledge of electronics is required to generate the UUT test program. For a functional board tester, however, a complete knowledge of the board is required. For most digital functional testing the engineer must manually generate the board's input test vectors to test all the logic on the UUT. In most cases, the test engineer has the aid of a logic simulator to help generate the UUT test program. A logic simulator consists of a computer with a large mass storage and a software package consisting of pre- and post-processors, a simulation algorithm, and a powerful editor. The pre-processor translates the UUT topography to interconnected library models resulting in a UUT model. The simulator will then promulgate these vectors through the board model first to initialize the UUT then to test for three classes of faults: stuck at 1, stuck at 0, and shorts. The simulator will store the resulting output test vectors, identify ICs input/output pins that have not been tested and display the percentage of faults that have been verified. The test engineer then manually generates additional input test vectors to test the IC pins that the simulator identified as not tested. This process continues until a QA rating is obtained that satisfies the test engineering manager. The post-processor software translates the test program generated by the simulator to operate on a particular functional board tester. It is not unusual for the digital test program generation time to be eight times as long for a functional board tester as for an in-circuit tester. If it takes one week to generate a digital test program for an in-circuit tester, it will take eight weeks to generate the test program for a functional board tester. This estimate is based on functional board test systems that employ algorithms to test the boards plus a simulator as a test program generation aid. According to a late 1983 Prime Data survey, test program generation for an in-circuit tester is accomplished 71 percent of the time on the in-circuit tester, 25 percent on a programming station, and the remaining 4 percent manually. For functional test program generation, 55 percent are performed on the functional board tester, 39 percent on a programming station, and 6 percent are done manually.

There are a couple of functional board comparison testers available that use quasi-random test vectors to stimulate both a known-good-board (KGB) and the board-under-test (UUT). This type of functional board testing is adequate as long as the KGB and UUT can be properly initialized and a defined sequence of test vectors is the testing source. Otherwise this test methodology is limited to SSI and some MSI technology. Program generation time for this type of comparison tester can take the same time as an in-circuit tester for a simple board containing SSI to 200 times longer for a complex board containing LSI.

Analog test program generation for a functional board tester requires the test engineer to create an entire decision tree to localize a fault. Thus the functional board tester's analog test program would take 20–30 times longer to generate the equivalent effective program for an in-circuit tester. The functional analog test program is a function of the skill and expertise of the programmer who must manually petition the analog and generate the individual diagnostic routines to localize failures. Aids for functional analog test programming are few and far between.

Debugging test programs for in-circuit testers is extremely fast because they deal with individual components. The debugging of a functional test program is comparatively slow. The actual debugging time required for a functional test program is essentially a direct function of knowledge and experience of the person doing the debugging.

Amending a test program to comply with an engineering change order (ECO) requires very little time with an in-circuit tester, while the same change order could take 2–4 times longer for the functional board tester depending upon how deeply the ECO effects the operation of the circuitry. If the ECO affects the artwork of the board, the in-circuit fixture test probe must be repositioned. Time is required to remove the old test probe, then drill, set, and wire a new test probe. This time may be the same as or even shorter than in a functional test system employing an edge connector interface.

The capital equipment cost for a functional board tester ranges from 1½ to 3 or more times the cost of an in-circuit tester. Functional board tester programming costs are from 4 to 20 times that for an in-circuit tester. The fixture cost for an in-circuit tester is 2–4 times that for a functional board tester. The assumption is that the functional board tester uses a type of edge connector and the in-circuit tester uses a bed-of-nails test fixture. However, a functional board tester employing a bed-of-nails fixture gets greater visibility and thereby reduces diagnostic time. Thus the cost of the functional fixture is about 75 percent of the cost of an in-circuit fixture. The overall cost of ownership over a 3–5 year period is very low for an in-circuit tester as compared to a functional board test system. It is not unusual to see cost ratios of 4 to 1, 10 to 1, or 14 to 1, depending upon the manufacturing facility. Generally the in-circuit and functional board test systems are equally reliable. Both have a mean time between failures (MTBF) on the order of 500 hours. Further the two systems have nearly the same mean time to repair (MTTR). For in-circuit testers the MTTR is typically one hour, while for the functional board testers, it is about two hours. The difference in MTTR is essentially the difference in the complexity of the test systems. However, converting either tester into a test system by employing a bed-of-nails fixture slightly reduces the reliability of the overall test system. This is due to the contributions of the higher failure rate items such as test probes and other mechanical portions of the fixture.

3.2. IN-CIRCUIT AND FUNCTIONAL BOARD TESTER SUMMARY

In summary, benefits of the in-circuit tester include:

1. Fast fault isolation and diagnostics
2. Test program generation requires lower technical level and less time
3. Shorter time required to integrate test system into production
4. Ability to prune marginal components that will pass a functional test
5. Ability to diagnose several failures in one pass
6. Low cost–performance ratio
7. Lower initial cost and cost of ownership

The functional board test system benefits include:

1. Tests unproven board designs
2. Tests component interaction
3. Detect intercomponent timing faults
4. Test boards with sensitive input/output signals
5. Test board's performance
6. Fast go/no-go test time

The disadvantages of an in-circuit tester include:

1. Will not test board performance
2. Will test at ambient temperatures only

3. Will not detect board timing faults
4. Will not test circuitry interaction
5. Slow go/no–go test time

The disadvantages of a functional board tester include:

1. Greater time and technical level required for generating the test programs
2. Cost of purchasing and ownership
3. Slow fault isolation and diagnostics

Many in-circuit testers have provisions to integrate three or four instruments into the system for limited functional testing. This capability has some effect on the above comparisons as the in-circuit tester would have some analog functional go/no–go capability in addition to the in-circuit diagnostic capability resulting in a limited combinational tester.

3.3. COMPARISON OF IN-CIRCUIT TESTER, IN-CIRCUIT ANALYZER, AND LOADED-BOARD SHORTS TESTER

Now that we have investigated and compared the advantages and disadvantages of in-circuit and functional board testers, we will begin a comparison between in-circuit testers (ICT), in-circuit analyzers (ICA), and loaded-board shorts testers (LBS). Table 3.2 is a list of topics with comparative responses for these types of testers.

As previously stated, in-circuit testing is a manufacturing verification tool. The in-circuit tester tests interconnections and individual components on a printed circuit board in both powered and unpowered states. By comparison, the in-circuit analyzer tests interconnections and individual components on a printed circuit board in the unpowered state. If the design of the board is proven, the active components have been pretested, and the artwork and passive components are correct, the board should be operational. The loaded-board shorts tester tests the interconnections on the printed circuit board. If the board design is proven and the artwork is correct, the board should be free of shorts and opens.

TABLE 3.2. In-Circuit/Analyzer/Shorts Tester Comparison.

Topic	In-Circuit Tester	In-Circuit Analyzer	Loaded-Board Shorts Tester
Type of testing	manufacturing verification	manufacturing verification	soldering verification
Test methodology	individual component	impedance signature	resistance crossover
Go/no-go test speed	slow	fast	fast
Diagnostic time	slow	fast	fast
Type of fault detection	manufacturing faults	shorts, R,C, semiconductor junctions	shorts/opens
Fault detection criteria			
Analog components	device specification	impedance signature	none
Digital components	pin faults	orientation and ID	none
Fault isolation	component level	component level	traces level
Fault resolution	component/net	component/net	traces/string
Fault accuracy	high	medium high	high
Fault detection capability	multiple	multiple	multiple
Test Hierarchy			
Unpowered PCB	2	2	1
Powered PCB	1 to 3	0	0
Logic timing detection	device level	none	none
Typical fault coverage			
Analog PCB	90–95%	70–92%	35–55%
Digital PCB	85–94%	50–75%	45–55%
Hybrid PCB	87–94%	60–87%	40–60%
Unverifiable faults	low	low	very low
Failure message	specific	specific/definable	specific

TABLE 3.2. In-Circuit/Analyzer/Shorts Tester Comparison. (continued)

Topic	In-Circuit Tester	In-Circuit Analyzer	Loaded-Board Shorts Tester
Technical levels required			
Test operator	low	low	low
Diagnostics operator	low	low	low
Programmer	medium	low	low
Maintenance	medium	medium	medium
Test programming effort			
Analog circuitry	low (1×)	low (0.4×)	low (0.1×)
Digital logic	low (1×)	low (0.2×)	low (0.1×)
Debug	low	very low	very low
Integration	medium	low	very low
Learning curve	3–6 mos.	1–2 mos.	1–3 wks.
Programming aides	APG	auto/self-learn	self-learn
100% utilization	4–9 mos.	1–2 mos.	2–4 wks.
ECO difficulty	low	low	very low
Capital cost	high	low	very low
Programming cost	medium	low	very low
Fixture cost	medium	medium	medium
Ownership cost	low	very low	very low
Reliability	high	high	high
MTBF	high	high	high
MTTR	low	low	low

An in-circuit tester electrically isolates the component/device-under-test (DUT) by guarding, then places the DUT into measurement circuitry to verify its specifications or logic performance. An in-circuit analyzer, on the other hand, obtains impedance signatures across two nodes, then compares these signatures with stored impedance signatures obtained from a known-good-board.

The loaded-board shorts testers typically force a voltage, measure a current, and then compare the results to selected resistance crossover values (10 ohms, 100 ohms, 1K, or 10K). The go/no-go test times and diagnostic test times are slower for the in-circuit tester than for either the in-circuit analyzer or the loaded-board shorts tester. This slowness is a result of the test methodology and speed of the switching matrix. The in-circuit analyzer and loaded-board shorts tester employ solid state switching, typically JFETS, and rapid threshold comparisons.

In-circuit testers have the ability to test for virtually all manufacturing faults, limited only by the design of the printed circuit board. The in-circuit analyzer tests for shorts, resistors, capacitance, and semiconductor junctions. Because of its test methodology, it has fewer components masked from detection by circuit board design. The loaded-board shorts tester tests for trace shorts and opens.

The in-circuit tester tests all passive components that are within the tester's measurement range to the device specifications, then powers up the board and, by employing a device library subroutines, stimulates and senses the digital logic elements for pin faults. The in-circuit analyzer tests all components to their stored impedance signature. The resistance signature consists of a measurement time delay to allow the voltage to be stable and one voltage sample. The capacitance signature consists of a discharge time, a measurement time delay, followed by three equally spaced voltage samples positioned on the voltage charge portion of the capacitor voltage/time response curve. The diode signature consists of two measurement time delays and two voltage samples. One voltage sample is taken in the forward bias direction and one voltage sample in the reversed bias direction. Both measurement time delays are to allow for stable voltage measurements. The transistor signature is the same as the diode signature except it consists of three combinations; emitter to base, emitter to collector, and base to collector, resulting in six signatures. IC orientation is a diode signature map from a power pin to all other pins, and from the ground pin to all other pins, generally limited to 20 signatures per IC.

The loaded-board shorts tester has no component fault detection capability.

Fault isolation for an in-circuit tester and an in-circuit analyzer is to the component level. As would be expected, the shorts tester is to the trace level. Fault resolution is normally definitive for all three types of in-circuit test systems. However, the in-circuit analyzer due to its test methodology, and especially if guarding is not employed, may list two or three possible faults in the order of their probability. Generally this occurs when a bad diode, transistor, or IC is invalid. The shorts fault resolution for all three types of test systems is identical.

All three types of in-circuit systems have multiple fault detection capability at each level of their test hierarchy. For unpowered PCBs, both the in-circuit tester and in-circuit analyzer have two levels of testing: shorts and components. Remember, the in-circuit analyzer component testing includes IC orientation. The shorts tester has only one level: shorts. For powered PCBs, the in-circuit tester has up to three levels: IC orientation, digital logic, and analog functions.

The typical fault coverage of an in-circuit tester is higher than that of an in-circuit analyzer because of the in-circuit tester's UUT power up test capability. The in-circuit tester can detect faults that the in-circuit analyzer cannot, and vice versa, as discussed in Chapter 4.

The unverifiable faults caused by system, operator, or UUT interface glitches are low in all instances. Failure messages from in-circuit test systems are generally specific. Even when the in-circuit analyzer produces a component net of 2–3 possible faults, the actual specific fault is generally definable.

For all three types of in-circuit testers only a moderately skilled technical level is required for testing and diagnosing PCBs. The in-circuit analyzer and the loaded-board shorts tester typically require less skill for test program generation because both systems have built-in software learn algorithms, whereby test programs are generated automatically from a known-good-board.

The test programming generation for an in-circuit analyzer is accomplished either by entering the generic parts and their nodal locations, then running an auto-learn routine on a known-good-board, or by placing a known-good-board on the tester, enabling auto-learn, entering the individual generic components, and probing their nodal locations. Shorts test program generation on the in-circuit analyzer is accomplished by placing a known-good-board on the tester and enabling the self-learn routine. For the shorts tester, the known-good-board is placed on the tester and the self-learn routine is enabled.

The debugging effort, based on the methodology of generating the initial test program, the in-circuit analyzers, and the shorts testers, is slightly less than that for in-circuit testers. Integrating the in-circuit tester into the production line requires adjustment of many individual tolerances based upon different component vendors. The in-circuit analyzer has an averaging algorithm which allows the final impedance signature to be determined by averaging values over a fixed number of PCBs. The shorts tester, requires no test program moderation as threshold levels are far below the range of component value variances.

The in-circuit tester has the longest learning curve: 3–6 months to the point where one feels comfortable with the test system. The loaded-board shorts tester has the shortest learning curve (1–3 weeks). The in-circuit analyzer is in the middle at 1–2 months. The amount of time required for 100 percent utilization is a function of the complexity of the test system, the complexity of the board-under-test, and the technical level of the user.

The capital investment for an in-circuit tester is in the range of $100,000–$300,000, the in-circuit analyzer $30,000–$40,000, and the shorts tester $15,000–$20,000. Programming costs are a function of the skill level of the programmer and the amount of time required to generate final test programs. The average programming time for a shorts tester is about 2–6 hours, for an in-circuit analyzer 1–4 days, and for an in-circuit tester 2–6 weeks.

Fixture costs are the same for all three types of in-circuit test systems. The cost of ownership for an in-circuit tester is higher than the cost for either an in-circuit analyzer or shorts tester. The reliability is high for all three types of test systems. The mean time between failure is about 700 hours and the mean time to repair is 1 hour.

3.4.   IN-CIRCUIT TESTER, IN-CIRCUIT ANALYZER, AND LOADED-BOARD SHORTS TESTER SUMMARY

In summary, comparing the in-circuit tester to the in-circuit analyzer, the in-circuit tester advantages include:

1. The ability to test IC logic functions
2. The ability to identify marginal components
3. Testing to specific values

4. Larger fault coverage
5. User acceptability

The in-circuit analyzer advantages include:

1. Faster fault isolation and diagnostics
2. Lower skill level and less time required for test programs generation
3. Faster integration into production
4. Faster product throughput
5. Lower cost of purchase and ownership
6. Lower cost performance ratio

The disadvantages of an in-circuit tester include:

1. Greater cost, time, and higher skill level required for test programs generation
2. Higher cost of purchase and ownership
3. Slower fault isolation and diagnostic time
4. Higher cost performance ratio

The disadvantages of an in-circuit analyzer include:

1. Limitations in fault coverage
2. Inability to detect IC logic functions
3. Inability to test powered PCBs
4. Limitations in fault resolution and accuracy

The advantages for the loaded-board shorts tester include:

1. Lowest cost of purchase and ownership
2. Fastest to integrate into production
3. Fastest test program generation
4. Lower skill level operator
5. Fastest fault isolation and diagnostics
6. Highest product throughput and cost performance ratio

The disadvantage of a loaded-board shorts tester is that it is limited to shorts and open testing.

3.5. LOW-COST ATE

What is low cost ATE? A de facto standard is ATE priced less than $100,000. As in everything we buy, we end up paying for what we get. However, with advances in technology, we get a lot more ATE performance for our dollar today than we did three years ago. But we usually don't buy what we need. We buy what we need plus a little more for insurance, security, prestige, or future anticipations. What are the tradeoffs between high- and low-cost ATE? Generally, the answer is far superior test comprehensiveness. In a specific case, the answer(s) may be faster test and diagnostic time, faster and more comprehensive test program generation, more complete data logging and analysis, and fewer requirements for a technically skilled operator. Are these answers a function of perception, reality, or the environment?

State-of-the-art ATE continues to increase in price. Because of the rising cost of meeting customers' needs of testing ICs with increased functional density and higher operating speeds, each new level of IC performance places still further demands on test preparation software and test execution hardware and software. The computer must have a capability to handle more instructions per second. In addition, the pin electronics includes gate arrays and completely custom devices. The software development effort to move with the leading edge of technology is enormous. Is this your dilemma, or is your testing requirement for proven, mature IC technology?

With the falling costs of computers, memory and peripherals – once the major ATE expense items – and advances in semiconductor technology, one should expect relatively high performance in low-cost ATE. Today's single-board computer with mass storage costs about $3,000; the personal computer with mass storage costs about $5,000; and the scientific minicomputer with mass storage costs well under $10,000. The advancement in hybrid technology enables a fairly complex circuit to go from design to production in about four months with a nonrecurring cost on the order of $15,000 to $30,000 and a recurring cost of $10 to $15 per unit. Gate array technology is comparable. Therefore, with current technology, low-cost ATE designed today should equal or surpass high-cost ATE designed three or four years ago.

Before discussing production ATE, let us consider for a moment the service department ATE. Service is considered a labor-intensive group with a very low capital equipment budget, typically employing a board-swapping strategy to minimize the user's system downtime, while optimizing customer satisfaction. In the majority of cases, PCBs rejected in the field are repaired at local field offices, centers, or depots rather than returned to manufacturing.

The service testing requirements are diversified because both old and new products must be diagnosed for faults, and repaired. To make these requirements effective, some companies have moved production ATE into field depots while other companies, in an attempt at frugality, have deployed low-cost ATE. Low-cost service ATE must satisfy testing needs and be flexible, reliable, and easy to program, all for less than $100,000.

Commercial vendors have responded to these service needs, offering in-circuit analyzers (ICA), in-circuit testers (ICT), functional board testers (FBT, both vector and in-circuit emulation), peripheral testers, power supply testers, and a variety of IEEE-488 rack and stack instrumentation. And while it is true all field-rejected PCBs were once totally operational and the service fault spectrum differs from that for production, the major difference between the two involves the UUT interface and the required operator's interaction. For service ATE, these result in a low throughput rate and a higher technical skill operator requirement as compared with production.

Service ATE is ideal for depot service, engineering prototype development, manufacturing pilots, and low-volume manufacturing.

3.5.1. Production Low-Cost ATE

Manufacturing defects associated with production tests do not appear to have undergone any major changes as a result of changes in component or packaging technology. In fact, many test engineers feel assembly- and soldering-induced faults are a greater concern. Both loaded-board shorts testers (LBST) and in-circuit analyzers are low-cost ATE in present use in many manufacturing facilities throughout the world.

3.5.1.1. Loaded-Board Shorts Tester. The advantages and disadvantages of a loaded-board shorts tester have been summarized earlier in this chapter.

Assuming shorts is the major fault mode, a low-cost loaded-board shorts tester, employed in tandem with a high-priced in-circuit tester – at little or no increase in production testing expense – will increase product throughput and increase the in-circuit tester's time availability for more complex testing. Further, an increase in product throughput, with a reduction in testing expense, is generally experienced when shorts are in excess of 55 percent of the fault spectrum or if the shorts test is deleted from the in-circuit tester's test hierarchy.

Typically, in an in-tandem configuration of a loaded-board shorts tester and an in-circuit tester, the loaded-board shorts tester is employed to detect shorts and the rework test verification is conducted by the in-circuit tester. The loaded-board shorts tester–in-circuit tester configuration usually increases product throughput by 5 to 20 percent at little or no additional expense, because after the initial startup, and concurrent in-circuit tester idle time while the pump is being primed, the first lot of PCBs passed by the loaded-board shorts tester are received by the in-circuit tester, and the test operation and repair are then conducted in parallel. As the loaded-board shorts tester is testing one group of PCBs at the rate of 5 to 7 per minute, the in-circuit tester is testing another group of PCBs at the rate of 2 to 4 per minute.

Eventually, the loaded-board shorts tester will run out of work and wait for the slower, high-priced in-circuit tester to finish the lot of PCBs. This parallel activity is true in all in-tandem configurations. The slowest testers determine the effective test rate and the product throughput.

When the loaded-board shorts tester and in-circuit tester are operated in tandem as separate entities, the total effective test is the same as stated above, with the further expense of the added loaded-board shorts tester PCB handling time and redundant defective PCB rework verification test time.

Deleting the shorts test from the in-circuit test hierarchy results in an overall reduction in the effective test rate, in turn resulting in an increase in throughput. Here the decrease in expense is realized because of a reduction in in-circuit tester usage, minus the additional handling time. The loaded-board shorts tester–in-circuit tester throughput should increase by a factor of 1.5 to 2.

When a loaded-board shorts tester is employed in tandem with a high-price functional board tester, the functional board tester's shorts test diagnostic requirements are offloaded to the loaded-board shorts tester, with a dramatic increase in product throughput and a significant reduction in testing expense.

The larger the number of shorts faults, the more the loaded-board shorts tester will benefit a production test facility. Unfortunately, the converse is also true.

The low-cost loaded-board shorts tester as a functional board tester screener results in a decrease in effective test time, and thus an increase in product throughput, compared to the use of a functional board tester alone. As a rule, the loaded-board shorts tester results in an increase in throughput of 5 to 10 times with a significant reduction in expense as compared to a functional board tester alone. Unfortunately, shorts typically are not the only failure mode.

3.5.1.2. In-Circuit Analyzer. In-circuit analyzer advantages and disadvantages were summarized earlier in this chapter.

Assuming manufacturing-induced faults and analog component faults are the majority of faults, a low-cost in-circuit analyzer employed in tandem with a high-priced in-circuit tester – at little or no increase in expense – will increase product throughput and increase the in-circuit tester's time availability for digital logic pin fault detection. An in-circuit analyzer will detect manufacturing-induced faults and analog component faults, excluding digital logic pin faults, generally with a third of the time and expense of an in-circuit tester. If manufacturing and analog faults exceed 95 percent of the fault spectrum, or if these tests (shorts, analog and digital orientation) are deleted from the in-circuit tester's test hierarchy, the in-circuit analyzer and in-circuit tester configuration results are an increase in product throughput with a reduction in testing expense compared to an in-circuit tester alone.

When a low-cost in-circuit analyzer is employed in tandem with a high-priced functional board tester, the functional board tester's test diagnostic requirements are for digital logic and performance faults. The in-circuit analyzer offloads the other fault modes, resulting in a dramatic increase in throughput at a significant reduction in expense as compared to a functional board tester alone.

3.5.1.3. Low Cost In-Circuit Testers. Both high- and low-cost in-circuit testers are presently used in production testing. Low-cost in-circuit testers are employed as stand-alone systems or implement test strategies in which testing simple PCBs cannot be cost-justified using a high-cost tester. Another test strategy is a master-slave or distribution system concept in which one high-cost tester is essentially the host (test-program storage, data analysis, and test-programming development) to a series of low-cost in-circuit testers.

The advantages of the low-cost in-circuit tester over its high-cost counterpart are:

- Lower initial capital investment, ranging from $50,000 to $100,000, typically $75,000
- Lower fault detection cost, typically less than $3 per fault
- Typically 20 percent less test cost than the high-cost in-circuit tester
- Higher MTBF and lower MTTR and cost of repair
- Lower cost of ownership.

The low-cost in-circuit tester's disadvantages are:

- Limited test program development capability
- Typically limited to low-speed digital testing
- Questionable data analysis capability
- Typically limited to 512 or 1,024 test point capacity
- Generally limited mass storage capability
- Slower total test time, typically 20 to 25 percent.

Essentially, the tradeoff between low-cost in-circuit testers and high-price in-circuit testers is a reduction in throughput for a reduction in total recurring expense. The general rule is that a 20 percent reduction in expense results in a 20 percent reduction in throughput.

3.5.1.4. Low-Cost Functional Board Tester. Both high- and low-cost functional board testers are presently used in production test. Low-cost functional board testers are employed as stand-alone systems or in implementing either a parallel test strategy as the go/no-go in-line tester or the testing of simple PCBs that cannot be cost-justified using a high-cost tester.

The advantages of a low-cost functional board tester as compared to its counterpart are:

- Lower testing cost
- Lower cost of ownership
- Cost-effective go/no-go capability
- Lower cost of fault detection, typically $6.25 per fault
- Lower initial investment, ranging from $50,000 to $100,000, typically $85,000.

Low-cost functional board tester disadvantages are:

- Test program development limited to editing capability
- No failute data analysis capability
- Limited test diagnostic capability
- Limited test point capability, typically less than 400 points
- Requires medium-skilled technician
- Limited mass-storage capability.

Generally, the low-cost functional board tester's test program is generated externally from a programming station or high-cost tester and ported to the low-cost tester.

The principal tradeoff between a low-cost functional board tester and a high-cost functional board tester is a lower initial cost and lower total recurring expense for a lower throughput. Generally, a 35-percent reduction in total expenses results in a 25-percent reduction in throughput.

3.5.1.5. Low-Cost Networking. A low cost local networking capability called IEEE 802.3, or ones even cheaper, is available. The IEEE 802.3 low-cost network operates at a 10-MHz data rate over thin, 0.25-inch diameter RG-58 coaxial cable terminated in standard BNC connectors. The 50-ohm system requires a transceiver IC to be installed inside each user's system and the data base is IEEE–488 format. This permits easy installation directly to the primary cable via a coaxial tee connection. The tradeoffs are a total cable span of 3,000 feet without a repeater and the limitation of the number of nodes to 30. For many production facilities, IEEE-802.3 may be worth investigation.

3.5.1.6. Low-Cost Tester Conclusions. The loaded-board shorts tester and the in-circuit analyzer presently existing in a larger number of production test strategies appear to be a very cost-effective solution to prescreening manufacturing-induced faults prior to functional board testing. Further, a loaded-board shorts tester, when shorts are the major fault, can be a cost-effective solution when screening with an in-circuit tester. An in-circuit analyzer, in a majority of cases, can also be a cost-effective solution when screening with an in-circuit tester. Both the loaded-board shorts tester and the in-circuit analyzer seem to be ideal investments for a first-time commercial ATE user.

Low-cost in-circuit testers have a definite place in today's production test world, performing cost-effective testing of small PCBs, either as stand-alone systems or in tandem with a functional board tester, or in a low-volume production and service facility in which throughput is not the major issue. The low-cost functional board tester has a defined place, relieving the high-cost functional board tester from testing simple PCBs, and in service facilities in which throughput is not the major issue.

The next-generation-design in-circuit testers and functional board testers should have a superior cost-performance ratio compared to present low-cost testers.

4
PCB PRODUCTION TEST STRATEGIES

The production test strategy and the selection of the proper test equipment to implement that test strategy is determined by first-pass-yield, fault spectrum, production volume, and board mix. An effective test strategy evolves concurrently with manufacturing efficiency and the company's three-year plan. The prime PCB test stations, test and rework, are loaded-board shorts testers (LBS), in-circuit analyzers (ICA), in-circuit testers (ICT), and functional board testers (FBT). By combining these prime test stations into various configurations, one can optimize production test capability. The first consideration is first-pass-yield.

4.1. TEST STRATEGIES

4.1.1. First-Pass-Yield

First-pass-yield is the number or percentage of PCBs that test good the first time with no rework involved. First-pass-yield is loosely categorized into high, medium, and low. High first-pass-yield is generally interpreted as in excess of 75 percent, medium first-pass-yield is considered between 40 and 75 percent, and low first-pass-yield below 40 percent. The production manager considers the product quality in terms of the type of test stations required to produce a fault-free product. The selection of the test station is strongly weighted by the capital cost, recurring cost, and most importantly, meeting the throughput or revenue objectives.

In the upper portion of the high first-pass-yield range, greater than 90 percent, which means the least number of faults that have to be diagnosed and reworked, the functional board tester with its fast go/no-go test time and high comprehensive testing capability is by far the best answer. However, as the first-pass-yield declines, the lengthy fault diagnostics of the functional board tester begins to erode the production throughput time. One strategy is to increase

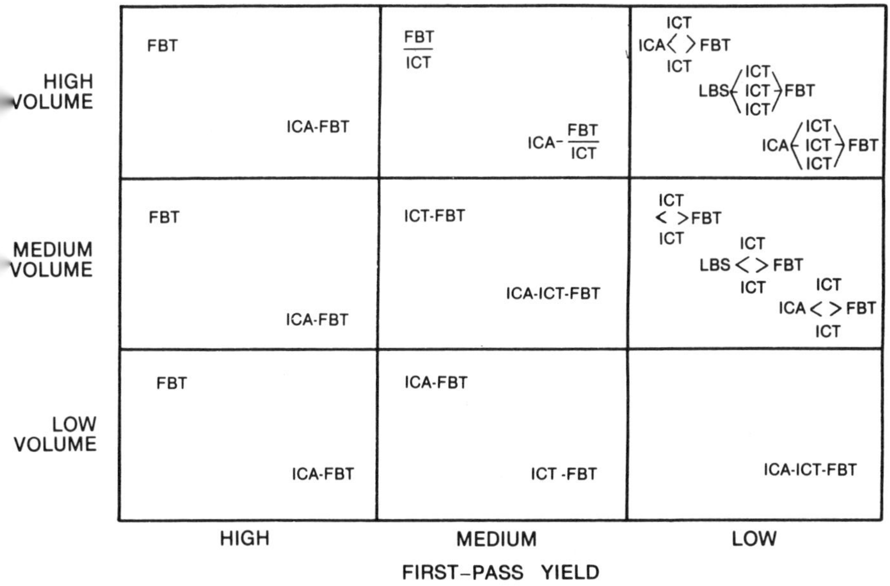

	HIGH	MEDIUM	LOW
HIGH VOLUME	FBT ICA-FBT	$\frac{FBT}{ICT}$ $ICA-\frac{FBT}{ICT}$	ICT $ICA\langle\ \rangle FBT$ ICT $LBS\{ICT\}FBT$ $ICA\{ICT\}FBT$
MEDIUM VOLUME	FBT ICA-FBT	ICT-FBT ICA-ICT-FBT	ICT $\langle\ \rangle FBT$ $ICT\quad ICT$ $LBS\langle\ \rangle FBT$ $ICT\qquad ICT$ $ICA\langle\ \rangle FBT$ ICT
LOW VOLUME	FBT ICA-FBT	ICA-FBT ICT-FBT	ICA-ICT-FBT

HIGH   **MEDIUM**   **LOW**

FIRST–PASS YIELD

Fig. 4.1. Test strategies.

the number of functional board testers and configure them in parallel.. This would be cost effective at the upper portion of the high first-pass-yield range. However, statistically, the predominant faults in high first-pass-yield category are assembly faults. This suggests that the most cost-effective solution would be to screen the functional board tester with an in-circuit analyzer (Fig. 4.1).

4.1.2. In-Tandem Testing

A screen or in-tandem test configuration is having test stations in series so the faults which a preceding tester is capable of detecting are identified and corrected before the product is presented to the next test station. The in-tandem test strategy is the most cost-effective method of detecting and reworking faults when maintaining a high level of throughput. The type of in-tandem test station that is employed in a test strategy is mainly determined by the fault spectrum. The first screening test station should have the highest diagnostic capability for the particular fault spectrum at the lowest recurring cost. The in-circuit analyzer in tandem with the functional board tester (ICA-FBT) in a high first-pass-yield situation results in the in-circuit analyzer test station diagnosing and reworking the majority of faults at the lowest recurring cost. The functional board tester diagnostic requirements would be reduced to a minimum. Further, due to the fast test and diagnostic capability of the in-circuit analyzer a higher throughput would be realized.

The test strategy for high first-pass-yield is essentially independent of PCB volume. For more throughput, in the first case, add another functional board tester in parallel. In the second case, first add another in-circuit analyzer in parallel. This should resolve the majority of issues. If not, add another functional board tester in parallel. The same test station configuration multiplied to meet the throughput demand. However for medium first-pass-yield situations because of the perceived longer diagnostic and rework times, the demand of product throughput or volume becomes a greater issue.

4.1.3. PCB Volume

Annual board volume is loosely defined as high, medium, or low. High-volume production is generally accepted as in excess of 750,000 PCBs per year, medium volume is between 250,000 and 750,000 PCBs per year, and low volume is below 250,000 PCBs per year. Annual PCB volume or revenue places a stronger influence on the test strategy to be employed. At the higher range of medium first-pass-yield, the predominant fault, in a large magnitude, is still assembly faults with shorts as a close second.

4.1.4. Parallel Testing

For high-volume and high-medium first-pass-yield situations the most cost-effective test strategy is a functional board tester in the main production line with an in-circuit tester in parallel. The functional board tester initially is in the role of a go/no-go functional board tester. The good PCBs pass down the production line and the rejected PCBs are immediately passed to the in-circuit tester to be diagnosed for manufacturing and component faults, as shown in Fig. 4.2(a).

IN-PARALLEL STRATEGY   IN-SHUNT TANDEM STRATEGY

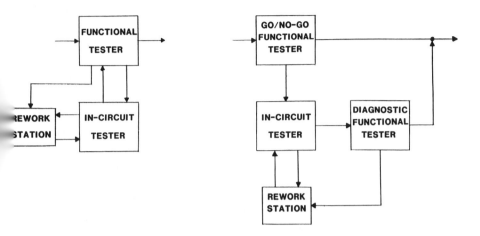

Fig. 4.2. Parallel test strategy.

The faults are then repaired and verified by the in-circuit tester. If good, the PCBs are passed onto the functional tester for performance testing. A performance failure is passed directly to the rework station for repair, verified by the in-circuit tester, and returned to the functional tester for performance testing.

The programming for the functional board tester is again simplified and focused on performance testing and diagnostics. The expensive functional and diagnostics rework time are reduced. The product throughput will typically increase 75–95 percent and the next station's yield will remain the same for PCBs that initially pass the functional test. PCBs with functional defects would have an increase in the next station's yield based upon the fault coverage of the in-circuit tester.

The modification to the in-parallel test strategy, called in-shunt tandem functional in-circuit configuration, Fig. 4.2(b), is a functional go/no-go tester in the production line, typically an in-house built system in parallel with an in-tandem in-circuit functional test configuration.

The go/no-go functional board tester optimizes the first-pass-yield and rejections are immediately passed to the in-circuit tester for identification of manufacturing and component faults. The faults are repaired and then verified by the in-circuit tester. If good, the PCBs are passed to the diagnostic functional board tester for performance testing.

The programming for the go/no-go functional board tester is very simple and no diagnostics are required. Programming for the diagnostic functional board tester is again simplified and focused on performance testing.

The expensive functional diagnostics and rework time are reduced and the increases in throughput and yield are the same for an in-parallel test strategy. The main advantage of an in-shunt tandem is better control and flow of the defective boards.

For medium-volume and high-medium first-pass-yield an in-circuit tester in tandem with a functional board tester appears to be the best cost-effective solution. The test systems are identical in capital cost to the high-volume test strategy, with the trade-off of high fault coverage with lower product throughput. The test configuration for the low-volume and high-medium first-pass-yield is an in-circuit analyzer in tandem with a functional board tester. This test strategy incurs less capital cost than the two previously stated test strategies, the trade-off being that more diagnostics are performed by the functional board tester.

In the low range of the medium first-pass-yield category, both the high and medium production test strategies are screened by an in-circuit analyzer. In the low-medium first-pass-yield category the predominant faults are shorts, followed very closely by assembly faults. In response, both the high- and medium-volume production test strategies screen their test station configuration with an in-circuit analyzer. Because of the in-circuit analyzer's test and diagnostic speed, no appreciable change occurs in throughput. The low-volume production screens the functional board testers with an in-circuit tester for the added diagnostic capability.

For low first-pass-yield all PCB production employs an in-tandem test strategy. High-volume production at 40–35 percent yield employs an in-circuit analyzer to screen multiple in-circuit testers in parallel which screens a functional board tester.

Generally speaking, an in-circuit analyzer has the capability of screening three in-circuit testers. Four or five in-circuit testers have the capability of screening one functional board tester. Two in-circuit analyzers have the capability of screening one functional board tester. Loaded-board shorts testers would be employed when the large majority of faults are shorts. A loaded-board shorts tester has the capability of screening one in-circuit analyzer, four in-circuit testers, or two functional board testers.

4.1.5. Board Mix

Another influence on the production PCB test strategy is the board mix. Generally the PCB board volume and different board types are inverse functions. High-volume production typically has a board mix of less than 20 types; medium-volume board mix generally ranges from 20 to 100 different types; and low-volume board mix is typically in excess of 100 different PCB types.

The degree of board mix affects setup time and cost. It also affects the manufacturing process first-pass-yield. Typically, production problems and board mix are in inverse proportion. Lower production volume and a large number of different board types affect the ability of the manufacturing process to be tuned for maximum performance and effective rework. The greater the board mix, the higher the setup costs and the lower the first-pass-yield and production throughput. The first-pass-yield for a product volume of greater than 30,000 PCBs per year is assumed to be 40 percent or lower for the higher board mixes, 40–75 percent for the medium board mixes, and greater than 75 percent for the lower board mixes.

Another factor in board mix is the complexity of the PCB. For many years various groups have unsuccessfully tried to define a standard PCB. The complexity of a PCB still remain subjective. The offering made by the American Association of Test Engineers is a PCB populated with 150 components, with a mixture of analog and MSI/SSI, is considered standard PCB. Boards with more than 150 components and/or LSI/VLSI devices are considered complex. The degree of complexity is assigned a numerical value by using a standard PCB multiplication factor. A microprocessor board is, for example, 2.8 times standard. This is a very broad definition. Perhaps not having a standard is best, as it forces detailed descriptions of the PCB and more accurate estimates should result.

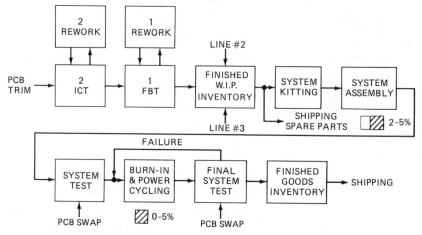

Fig. 4.3. Medium-volume PCB test

4.2. MEDIUM-VOLUME TEST STRATEGY

Figure 4.3 shows production test strategy for a medium-volume computer peripheral manufactured in the Sunbelt with an annual volume of 500,000 boards per year. The finished product has a board mix of six with a total complement of 22 boards per unit. The PCBs are tested initially on two in-circuit testers. If they fail, they are reworked and retested on an in-circuit tester. Upon passing the in-circuit test, they are routed to a functional board tester. Upon passing the functional test, they are placed in finished work-in-process (WIP) inventory. Upon demand, a system is kitted from the WIP inventory and routed to system assembly. After assembly, system test is performed, followed by 24-hour burn-in with a one-hour power cycling, and then the board is routed to final test. Upon passing final test, the system is packaged and held in finished goods inventory for shipment.

4.3. HIGH-VOLUME TEST STRATEGY

Figure 4.4 illustrates a production test strategy of a high-volume Midwestern manufacturer producing a single board product. The product is a microprocessor-based controller board. It has analog inputs that are digitized and digitally manipulated by a microprocessor,

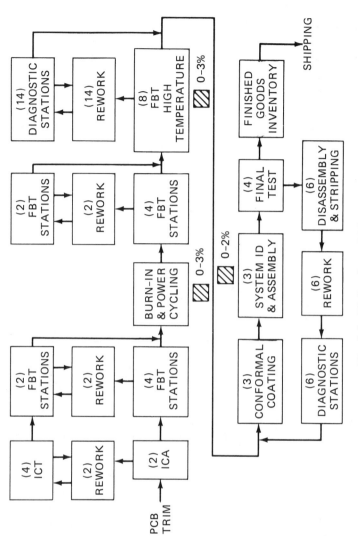

Fig. 4.4. High-volume PCB test.

preprogrammed by a PROM, and then converted back to analog control signals. The first-pass-yield is 89 percent with a throughput rate of a PCB every 15 seconds.

The PCBs from solder/wash/trim are first tested on an in-circuit analyzer. If good, the boards pass to an in-house-built functional board tester. If bad, the defective PCBs are routed to a rework station. After rework, the repaired PCB is verified on an in-circuit tester. If more defects are found, the faulty PCB is routed back to rework, and after repair it is again retested on the in-circuit tester. When the PCB passes the in-circuit tester, it is passed to an in-house functional board tester. If a fault is detected, the faulty board is sent to a rework station, and upon repair it is retested on the functional board tester. If passed, it is installed in the burn-in ovens for a 24-hour period at 125 degrees Farenheit, with 1-hour power cycling. After the baked boards are tested on an in-house-built functional board tester.

If a defect is found, the board is routed to a rework station and, after repair, is routed to a functional board test station. Upon passing, the board is routed to one of eight high-temperature soak ovens that has provisions for testing the PCBs at high temperature. If a failure occurs, the defective PCB is repaired at a rework station and passed to one of 14 diagnostic stations. Each diagnostic station has an in-house-built simulator that can recreate the operating environment of control boards.

Upon passing all diagnostic tests, the board is conformally coated and passed on to the system assembly area where a prom is installed to identify the specific operation of the control unit. It is then passed on to final test. If the PCB passes final test, it is placed in finished goods inventory and is ready for shipment. If the PCB fails final test, the prom is removed and the conformal coating is stripped. The board is then reworked and verified in a diagnostic test station. Upon passing, the board will be conformally coated again and routed through the remainder of the test cycle. A board is scrapped if it fails a specific level of tests three times in a row.

Having the functional board tester before burn-in and after appears to be redundant testing. The ambient functional board tester's fault detection rate is 1 percent and the post-burn-in functional board tester's fault detection rate is 2 percent. By removing the first functional board tester a reduction in direct labor cost and an increase in throughput will be realized without affecting the product quality.

4.4. PCB TEST STRATEGY EVALUATION

To obtain a better insight to the PCB production test strategies, let us take as an example the fault spectrum of 1000 disk-drive controllers discussed in Chapter 1. This manufacturing process produced a minimum first-pass-yield of 60 percent with an average of 0.5 faults per board, or an average of 1.3 faults per faulty board. The actual fault summary was illustrated in Table 1.2. The first-pass-yield is 602 boards with 398 defective boards containing 512 faults.

4.4.1. Fault Coverage

Table 4.1 illustrates the fault coverage for the four prime PCB testers. Of the 512 total faults, the loaded-board shorts tester detected a total of 266 faults for a coverage of 52 percent. It also rejected 223 out of the 398 PCBs as containing shorts. After the shorts are cleared in re-work, 195 PCBs are free of faults. Of the 1000-PC-board lot tested, 203 boards still had undetected faults. Similarly, the in-circuit analyzer detected a total of 421 faults for a fault coverage of 82 percent. A total of 345 PCBs was rejected, of which the rework yield is 316. Of the total lot tested and repaired, 82 boards had undetected faults.

The in-circuit tester detected a total of 462 faults for a fault coverage of 90 percent. A total of 370 PCBs were rejected, of which the rework yield was 354. Of the total lot tested, 44 boards had undetected faults.

The functional board tester detected a total of 486 faults for a fault coverage of 95 percent. A total of 385 PCBs were rejected, which the rework yield is 376 PCBs. Therefore, of the total lot tested, 22 boards had undetected faults.

Table 4.2 illustrates another insight into the actual faults detected by the in-circuit analyzer, the in-circuit tester, and the functional board tester. The dots indicate undetected faults. Consider the in-circuit analyzer fault detection at one fault per PCB (second column). The first fault indicates 26 wrong analog components and the in-circuit analyzer detected all 26. Continuing across the table, the in-circuit tester detected 25 and the functional board tester detected 23. On investigating the fault coverage of each of these three prime systems, one can see that, because of the different test methodologies, one type of tester would detect faults that another type would fail to detect.

TABLE 4.1. Tester's Fault Coverage.

Fault Classification	LBS				ICA				ICT				FBT			
	1	2	3	4	1	2	3	4	1	2	3	4	1	2	3	4
Shorts (S)	156	79	21	5	156	79	21	5	156	79	21	5	156	79	21	5
Opens (O)	2	2	1	0	2	2	1	0	2	2	1	0	2	2	1	0
Missing Components																
Analog (MA)					11	5	1	1	11	7	2	1	10	6	1	1
Digital (MD)					4	2	1	0	4	2	1	0	4	2	1	0
Wrong Components																
Analog (WA)					26	16	0	1	25	17	1	2	23	15	0	2
Digital (WD)					2	7	1	0	8	7	1	0	8	7	1	0
Misorientation																
Analog (RA)					11	3	1	1	11	3	0	1	10	3	0	1
Digital (RD)					8	1	0	1	7	0	0	1	9	1	0	1
Bent Leads																
Analog (BA)					21	4	1	0	21	4	2	0	21	4	2	0
Digital (BD)					6	4	1	1	10	4	1	1	10	4	1	1
Analog Specs (AS)					10	2	1	0	14	6	1	0	13	4	1	0
Digital Logic (DL)									11	6	2	1	14	8	3	2
Performance																
Analog (PA)													6	3	1	1
Digital (PD)													8	5	2	0
Total No. of Faults	158	81	22	5	257	125	29	10	280	137	33	12	294	143	35	14
Rejected PCBs	158	48	13	4	257	71	13	4	280	73	13	4	294	74	13	4
Test/Rework Yield	158	33	4	0	257	54	5	0	280	64	8	2	294	69	10	3

TABLE 4.2. Testers Fault Detection.

ICA

1 Fault Per PCB

156 S	26 WA	21 BA ●	DL ●
2 O	● WD 2	● BD 6	PA ●
● MA 11	11 RA	● AS 10	PD ●
4 MD	● RD 8		

2 Faults Per PCB

31 S-S	1 S-●	2 WA-AS ●	3 WA-●
2 S-O	2 S-●	1 ●-RD	1 BD ●
3 S-MA	3 S-●	1 BA-BD	1 ●-●
3 S-WA	7 WA-WD	1 ●-●	
2 S-RA	1 ●-RA	2 MA-MD	1 BD ●
1 S-BD	3 ●-BA	1 ●-WA	2 ●-●

3 Faults Per PCB

3 S-S-S	1 S-●-RA ●
1 S-S-O	1 S-MD-BD ●
1 S-S-●	1 S-●-AS ●
1 S-MA-●	1 S-●

4 Faults Per PCB

| 1 S-S-●-BD | 1 S-●-RA ● |
| 1 S-WA-MA-DL | 1 S-●-RD ● |

ICT

1 Fault Per PCB

156 S	● WA 25	21 BA ●	DL 11 ●
2 O	8 WD	10 BD ●	PA ●
● MA 11	11 RA	● AS 14	PD ●
4 MD	● RD 7		

2 Faults Per PCB

31 S-S	1 S-AS	2 ●-AS	3 WA-DL
2 S-O	2 S-●	1 ●-RD	1 BD ●
3 S-MA	3 S-●	1 BA-BD	1 MA-DL
3 S-WA	7 WA-WD	1 AS-AS	1 AS ●
2 S-RA	1 ●-●	2 MA-MD	1 BD-●
1 S-BD	3 WA-BA	1 MA-WA	2 ●-●

3 Faults Per PCB

3 S-S-S	1 S-●-●	1 S-BA-●
1 S-S-O	1 S-MD-BD	1 S-WD-●
1 S-S-MA	1 S-BA-AS	1 S-DL-●
1 S-MA-WA	1 S-DL-●	

4 Faults Per PCB

| 1 S-S-WA-BD | 1 S-MA-RA-DL ● |
| 1 S-●-● | 1 S-WA-RD-● |

FBT

1 Fault Per PCB

156 S	● WA 23	21 BA ●	14 DL
2 O	8 WD	10 BD ●	6 PA
● MA 10	● RA 10	● AS 13	8 PD
4 MD	9 RD		

2 Faults Per PCB

31 S-S	1 S-AS	2 ●-●	3 WA-DL
2 S-O	2 S-PD	1 RA-RD	1 BD-DL
3 S-MA	3 S-DL	1 BA-BD	1 ●-DL
3 S-WA	7 WA-WD	1 AS-AS	1 AS-PA
2 S-RA	1 WA-●	2 MA-MD	1 BD-PD
1 S-BD	3 ●-BA	1 MA-WA	2 PD-PA

3 Faults Per PCB

3 S-S-S	1 S-●-●	1 S-BA-PA
1 S-S-O	1 S-MD-BD	1 S-WD-PD
1 S-●	1 S-BA-AS	1 S-DL-PD
1 S-MA-●	1 S-DL-DL	

4 Faults Per PCB

| 1 S-S-WA-BD | 1 S-MA-RA-DL |
| 1 S-●-DL | 1 S-WA-RD-PD |

4.4.2. In-Tandem Testing Evaluation

The in-circuit analyzer and the in-circuit tester each detected 413 identical faults for a fault coverage of 81 percent. Further, the in-circuit analyzer detected 8 faults that the in-circuit tester missed, and the in--circuit tester detected 48 faults that the in-circuit analyzer missed. If these two systems are used in an in-tandem configuration, the total fault coverage would be 470 faults or 92 percent. That is an increase in fault coverage of 2.2 percent over the in-circuit tester alone.

The in-circuit analyzer and the functional board tester both detected 407 identical faults. The in-circuit analyzer detected 14 different faults that the functional board tester missed, and the functional board tester detected 76 faults that the in-circuit analyzer missed. If the in-circuit analyzer and the functional board tester are placed in an in-tandem configuration, the total fault coverage would be 497 faults or 97 percent. That is an increase in fault coverage of 2.3 percent over the functional board tester alone.

The in-circuit tester and the functional board tester, both detected 449 identical faults for a fault coverage of 88 percent. The in-circuit tester detected 13 faults that the functional board tester missed, and the functional board tester detected 36 faults the in-circuit tester missed. If the in-circuit tester and the functional board tester were in an in-tandem configuration, the total fault coverage would be 498 faults or 97 percent. That is a 2.5 percent increase in fault coverage over the functional board tester alone.

Conversely, if the in-circuit analyzer and the in-circuit tester and the functional board tester were placed in three-level, in-tandem test strategy, the total fault coverage would be 508 or 99 percent. That would be a 4.5 percent increase over the functional board tester alone, or an increase of 2 percent over the in-circuit tester and functional board tester in-tandem.

Table 4.3 lists the fault coverage for in-tandem configurations of the three prime PC board testers. Of the 398 defective PCBs, note the number of PCBs rejected by the various test strategies and the rework yield for each of the test strategies. Thus, the number of undetected faulty PCBs passed to the next test station will be 37, 14, 10, and 5, respectively.

Figure 4.5 illustrates the typical fault coverage increases for the in-tandem test strategies based upon the type of PCBs being tested.

TABLE 4.3. In-Tandem Fault Coverage.

FAULT CLASSIFICATION	ACTUAL	ICA-ICT	ICA-FBT	ICT-FBT	ICA-ICT-FBT
Shorts	261	261	261	261	261
Opens	5	5	5	5	5
Missing components	30	29	27	29	30
Wrong components	67	64	59	60	65
Reversed components	28	27	28	25	28
Bent leads	43	43	43	43	43
Analog specifications	25	21	21	22	23
Digital logic	27	20	27	27	27
Performance	26		26	26	26
Total No. of Faults	512	470	497	498	508
Fault coverage	100%	92%	97%	97%	99%
Rejected PCBs	398	374	391	393	394
Rework yield	378	361	384	388	393

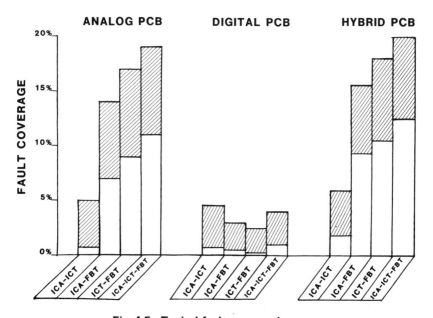

Fig. 4.5. Typical fault coverage increase.

Note that the largest range increases are for analog PCBs. When screening a functional board tester for digital PCBs, the amount of fault coverage increase is generally less than 5 percent. An in-circuit analyzer (ICA) screening an in-circuit tester (ICT) increases the capability of detecting misoriented ICs because the ICA takes a footprint of the semiconductor junctions of an unpowered IC and a toeprint of the resistance between specific IC pins. This is more accurate than powering the IC and pulsing each lead high and low to determine the I/O pattern of the IC, as would be necessary with the ICT.

An in-circuit tester screening a functional board tester (FBT) has a large duplication in fault coverage because the ICT tests for pin faults such as stuck at 1, stuck at 0, adjacent pin shorts, and pin-to-pin shorts. The added fault coverage is based upon failing IC output voltages that, although below vendor specifications, perform logically in the circuit. Further, output voltage degradation would cause a functional failure.

The in-circuit analyzer, in-circuit tester, and functional board tester in tandem would provide optimum fault coverage independent of the type of PC board being tested. As suspected, in the hybrid PCB arena, the amount of increase in fault coverage by in-tandem test strategy is essentially a function of the mix between analog and digital components on the board.

In-circuit testers will identify manufacturing and component faults for fast rework and verify the rework before functional testing is performed. This screening helps overcome the greatest handicap of functional testing by enabling the programming for the functional board tester to be simplified and focused on performance faults. In addition, the expensive functional diagnostic time is sharply reduced. Finally, this arrangement cuts rework time. The anticipated increase in product throughput would be 60–86 percent. Obviously, the capital cost of an in-circuit tester plus a functional board tester is less than the capital cost of two functional board testers.

Consider the test strategy that places an in-circuit analyzer in front of a functional board tester. The in-circuit analyzer tests unpowdered boards and would detect manufacturing and component faults. However, due to the fault coverage as compared to an in-circuit tester, the anticipated increase in throughput is 15–40 percent.

In another possible strategy, a loaded-board shorts tester screens a functional board tester. The shorts tester identifies all shorts and, after repair, verifies that the PCB is free of shorts. The throughput increase is anticipated to be 10–25 percent. The yield is unchanged. The cost/performance ratio is better than with the screened in-circuit tester combination, the cost being a fourth of an in-circuit tester. However, an in-circuit tester would give a 60–85 percent increase in capacity plus an increase in yield.

When shorts are the predominant PCB fault, one should consider placing a loaded-board shorts tester in front of an in-circuit tester. The testing time is typically 800–900 microseconds per test point. In many cases, this is faster than in-circuit tester performance. The loaded-board shorts tester identifies all independent shorts on the UUT in a single test. The board is then passed to the in-circuit tester which immediately makes its tests for shorts. Finding none, it would continue into its test hierarchy. The increased throughput is typically 5–12 percent and the product yield is unchanged. The cost/performance ratio in this combination is questionable because for approximately 50 percent more, an in-circuit analyzer could be purchased in place of a loaded-board shorts tester. This would increase both fault coverage and throughput. However, a serious shorts problem and a tight budget allotment might make the in-tandem arrangement more attractive because of the increase in throughput alone.

4.4.3. Test and Rework Time

Table 4.4 illustrates the PCB example of the 1000-controller-board test and rework time. The top half of the table gives average test time in minutes. As stated previously, for in-circuit testers the sum of the diagnostic times is the good PCB test time. The loaded-board shorts tester can, having only one hierarchy, detect all the shorts on the PCB in the time required to define the board as being good. The in-circuit analyzer has two levels of test hierarchy; the in-circuit tester has four levels.

For the functional board tester, the PCB test-good time is far shorter than any of the diagnostic times, and each diagnostic test is to identify a single fault. The PC board handling time is defined as 0.251 minutes for all the prime PC board test systems.

TABLE 4.4. Test and Rework Time.

Test Function	Average Test Time in Minutes			
	LBS	ICA	ICT	FBT[a]
Good PCBs	0.033	0.100	0.333	0.050
Shorts diagnostics	0.033	0.033	0.083	5.18/F
Analog diagnostics		0.100	0.217	6.50/F
IC orientation		INCL	0.218	INCL
Digital Diagnostics			0.333	3.84/F
PCB Handling			0.251	0.251
Rework Function	Average Rework Time in Minutes			
Shorts and opens	3.41	3.41	3.41	4.73
Analog components		1.94	1.58	3.12
Digital components		1.73	1.61	2.67
Tote transport	5.55	5.55	5.55	5.55

[a] /F = per fault.

The lower half of Table 4.4 shows the average rework time in minutes. Again this is the time required to read the failure message, localize the defect on the PCB, and make repairs. The amount of time required to transport the PCBs from the tester to the rework station is defined as 5.55 minutes.

Employing this data base for an example of 1000 control boards, Tables 4.5(a) and 4.5(b) illustrate calculations for an in-circuit and a functional board station, respectively. The in-circuit test system on its first lot test identifies 223 PCBs containing shorts. At the rework station, 266 shorts were repaired. Conversely, on the functional board tester, on the first lot test, 223 PCBs were defined as having shorts. The rework station repaired the same number, 223. This is due to the multi-fault detection capability of the in-circuit tester versus the single fault detection of a functional board tester.

The results of these calculations and similar calculations for the loaded-board shorts tester and in-circuit analyzer are shown on Table 4.6. The fault coverage from Chapter 3 is indicated on the first line. The calculated test-good time clearly indicates that the functional board shorts tester, at 0.83 hours, is extremely fast for the degree of test thoroughness and comprehensiveness. Conversely, the in-circuit tester, at 5.55 hours testing time, is considerably slower

Test good   630 × 0.333 = 3.497 hrs
Test shorts   223 × 0.083 = 0.308 hrs
Test analog   104 × 0.217 = 0.376 hrs
IC orientation   8 × 0.218 = 0.029 hrs
Test digital   35 × 0.333 = 0.194 hrs
PCB handling   1000 × 0.251 = 4.183 hrs

Rework shorts   266 × 3.41 = 15.118 hrs
Rework analog   109 × 1.58 = 2.870 hrs
Rework digital   43 × 1.61 = 1.154 hrs
Tote transport   2 × 5.55 = 0.185 hrs

Test good   335 × 0.333 = 1.859 hrs
Test analog   16 × 0.217 = 0.058 hrs
Test digital   19 × 0.333 = 0.105 hrs
PCB handling   370 × 0.251 = 1.548 hrs

Rework analog   19 × 1.58 = 0.500 hrs
Rework digital   20 × 1.61 = 0.537 hrs
Tote transport   2 × 5.55 = 0.185 hrs

Test good   32 × 0.333 = 0.178 hrs
IC orientation   1 × 0.218 = 0.004 hrs
Test digital   2 × 0.333 = 0.011 hrs
PCB handling   35 × 0.251 = 0.146 hrs

Rework digital   3 × 0.333 = 0.081 hrs
Tote transport   2 × 5.55 = 0.185 hrs

Test good   3 × 0.333 = 0.017 hrs
PCB handling   3 × 0.251 = 0.013 hrs

Test good   615 × 0.050 = 0.513 hrs
Test shorts   223 × 5.167 = 19.204 hrs
Test analog   104 × 6.500 = 11.267 hrs
Test digital   58 × 3.94 = 3.809 hrs
PCB handling   1000 × 0.251 = 4.183 hrs

Rework shorts   223 × 4.93 = 18.323 hrs
Rework analog   104 × 3.12 = 5.408 hrs
Rework digital   58 × 2.67 = 2.581 hrs
Tote transport   2 × 5.55 = 0.185 hrs

Test good   300 × 0.050 = 0.250 hrs
Test shorts   39 × 5.167 = 3.359 hrs
Test analog   19 × 6.500 = 2.058 hrs
Test digital   27 × 3.940 = 1.773 hrs
PCB handling   386 × 0.251 = 1.615 hrs

Rework shorts   39 × 4.93 = 3.205 hrs
Rework analog   19 × 3.12 = 0.988 hrs
Rework digital   27 × 2.67 = 1.202 hrs
Tote transport   2 × 5.55 = 0.185 hrs

Test good   72 × 0.050 = 0.060 hrs
Test shorts   4 × 5.165 = 0.344 hrs
Test analog   4 × 6.500 = 0.433 hrs
Test digital   5 × 3.94 = 0.328 hrs
PCB handling   85 × 0.251 = 0.356 hrs

Rework shorts   4 × 4.93 = 0.329 hrs
Rework analog   4 × 3.12 = 0.208 hrs
Rework digital   5 × 2.67 = 0.223 hrs
Tote transport   2 × 5.55 = 0.185 hrs

Test good   10 × 0.050 = 0.008 hrs
Test digital   3 × 3.94 = 0.197 hrs
PCB handling   13 × 0.251 = 0.054 hrs

Rework digital   3 × 2.67 = 0.134 hrs
Tote transport   2 × 5.55 = 0.185 hrs

Test good   3 × 0.050 = 0.003 hrs
PCB handling   3 × 0.251 = 0.013 hrs

TABLE 4.6. PCB Tester Comparison.

PRIME PCB TESTERS

→	LBS	→ →	ICA	→ →	ICT	→ →	FBT	→
	↑↓		↑↓		↑↓		↑↓	
	REWORK		REWORK		REWORK		REWORK	

Fault coverage	52%	82%	90%	85%
Test good time (hrs)	0.55	1.67	5.55	0.83
Diagnostic time (hrs)	0.12	0.36	1.09	42.77
PCB handling time (hrs)	5.12	5.71	5.89	6.22
Rework time (hrs)	15.12	20.02	20.26	32.60
Tote transport (hrs)	0.19	0.37	0.56	0.74
Test time/PCB (min)	0.35	0.46	0.75	2.99
Throughput/PCB (min)	1.27	1.69	2.00	4.99
Process yield	80.5%	91.8%	95.6%	97.8%

than any of the other prime PC board testers. However, in looking at the diagnostic time, the in-circuit tester performs its diagnostics in 1.09 hours, while the functional board tester requires 42.77 hours. That is an extremely long diagnostic time as compared to in-circuit systems.

Printed circuit board handling time reflects the number of retests required to define all the faults the test system is capable of detecting. Note the rework time for the in-circuit analyzer and the in-circuit tester are essentially the same. However, the functional board tester, because of the requirement of interpreting its failure message, adds hours to the processing with only a very small increase in fault coverage.

The tote time reflects the number of times PCBs are transported back and forth between the rework station and the tester. The average test time per PCB, in the case of an in-circuit tester at 0.75 minutes, is a reflection of the test-good time. For the functional board tester, the average test time per PCB of 2.99 minutes is a reflection of the diagnostic time.

TABLE 4.7. PCB Test Strategy — In-Circuit.

SCREENED IN-CIRCUIT TESTER

```
 → ┌──────┐  ┌──────┐      ┌──────┐    ┌──────┐
   │ LBS  ├─→│ ICT  ├─ → → │ ICA  ├──→ │ ICT  ├─→
   └──────┘↑ └──────┘      └──────┘    └──────┘
      │  │    ↑   ↑          ↑  ↑        ↑  ↑
 ┌──────┐  ┌──────┐      ┌──────┐    ┌──────┐
 │REWORK├─ │REWORK│      │REWORK│    │REWORK│
 └──────┘  └──────┘      └──────┘    └──────┘
```

Fault coverage	90%	92%
Test good time (hrs)	4.76	6.90/7.22
Diagnostic time (hrs)	1.74	0.61
PCB handling time (hrs)	9.05	9.90/9.98
Rework time (hrs)	12.54/20.10	11.80/21.69
Tote transport (hrs)	0.83	1.30/1.39
Test time/PCB (min)	0.93	1.04
Throughput/PCB (min)	1.73	1.83
Process yield	95.6%	96.3%

The average throughput per PCB (two minutes for an in-circuit tester) essentially reflects the test good time and the rework time. For the functional board tester, 4.99 minutes reflects the diagnostic time and rework time. The process yield from the previous chapter is stated on the last line as a reference of accomplishment. The in-circuit tester station took 33.35 hours to obtain a 95.6 percent process yield, whereas the functional board tester took 83.16 hours to obtain a 97.8 percent process yield.

Table 4.7 shows results of in-tandem testing configuration of an in-circuit tester screened by a loaded-board shorts tester (LBS-ICT), then an in-circuit analyzer (ICA-ICT). Whenever two or more test systems are configured in tandem, some activities are occurring simultaneously. One of these activities has an immediate effect on the product throughput while the other activity does not. The time in the upper portion of the diagonal affects throughput, while the time in the bottom portion of the diagonal is the total active time and does not affect throughput. The fault coverage is on the first line. There is no increase in fault coverage by screening an in-circuit tester with a loaded-board shorts tester, as both these systems adequately cover shorts. There is, however, as discussed in the previous chapter, an increase in fault coverage when an in-circuit tester is screened by an in-circuit analyzer (ICA-ICT).

The test-good time of a loaded-board shorts tester and an in-circuit tester in tandem is 4.76 hours, while the sum of the two times is 6.1 hours. The reduction in time is due to the confirmation of reworked PCBs being conducted by the in-circuit tester rather than by the loaded-board shorts tester. For the in-circuit analyzer in tandem with the in-circuit tester, the test-good time totals 7.22 hours. However, because operations are conducted in parallel, the time that affects throughput is 6.9 hours.

For the LBS-ICT, as compared to an ICT, there is a decrease in test-good time of 0.79 hours, whereas for the ICA-ICT, there is an increase of 1.35 hours. Conversely, the diagnostic time of the LBS-ICT increases by 0.65 hours, and the ICA-ICT decreases by 0.48 hours.

The PC board handling time, as compared to an ICT, increases by 3.16 hours for the LBS-ICT and 4.01 hours for the ICA-ICT. Both LBS-ICT and ICA-ICT configurations show a reduction in rework time. However, there is an increase in total transportation time and in average test time per PCB. The result is product throughput increases as the throughput time per PCB decreases in the case of an LBS-ICT from two minutes to 1.73 minutes and for an ICA-ICT from two minutes to 1.83 minutes.

A dramatic change in test time per PCB and throughput time per PCB occurs when a functional board tester is screened as illustrated in Table 4.8. The test time per PCB employing a functional tester alone is 2.99 minutes; however, this decreases rapidly to 1.12 minutes when the tester is screened by either an ICA or an ICT. The test time decreases further to 1.05 minutes when screened by the combination of an ICA-ICT. The throughput time per PCB decreases from 4.99 minutes to approximately 2 minutes for each of the screening configurations.

4.4.4. Recurring Costs

The top portion of Table 4.9 gives the labor rates employed for the various types of test equipment. The difference in labor rates is

TABLE 4.8. PCB Test Strategy — Functional.

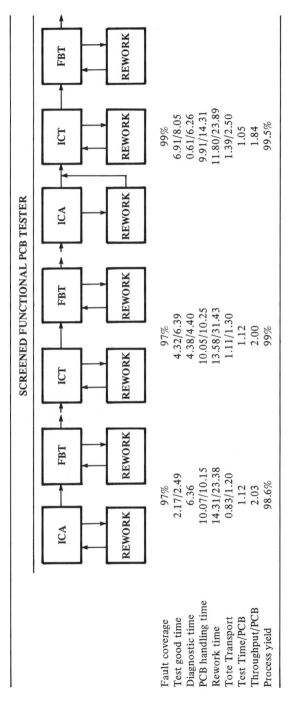

SCREENED FUNCTIONAL PCB TESTER

	ICA		ICT		ICA	ICT
Fault coverage	97%	97%				99%
Test good time	2.17/2.49	4.32/6.39				6.91/8.05
Diagnostic time	6.36	4.38/4.40				0.61/6.26
PCB handling time	10.07/10.15	10.05/10.25				9.91/14.31
Rework time	14.31/23.38	13.58/31.43				11.80/23.89
Tote Transport	0.83/1.20	1.11/1.30				1.39/2.50
Test Time/PCB	1.12	1.12				1.05
Throughput/PCB	2.03	2.00				1.84
Process yield	98.6%	99%				99.5%

TABLE 4.9. Example Costing.

	LBS	ICA	ICT	FBT
Operation	Typical Labor Rates per Hour (unloaded)			
Programming	$ 6.00	$10.00	$11.00	$18.00
Testing	6.00	6.00	6.00	10.00
Rework	6.00	6.00	6.00	12.00
Toting	4.00	4.00	4.00	4.00
Cost Centers	Typical Fixed Cost Amortization			
System(s) cost	$26,000	$45,000	$187,000	$276,000
Amortized − 5 yr/work hour	2.58	5.49	22.48	33.68
Fixture	1,412	1,412	1,412	287
Programming	51	215	1,220	6,063
System time	9	76	1,803	9,913
Total setup cost	1,472	1,703	4,435	16,263
Amortized/12,000 PCBs	$ 0.12	$ 0.14	$ 0.37	$ 1.36

TABLE 4.10. Recurring Cost — Prime.

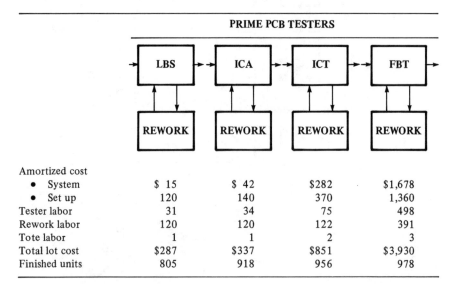

PRIME PCB TESTERS

Amortized cost				
• System	$ 15	$ 42	$282	$1,678
• Set up	120	140	370	1,360
Tester labor	31	34	75	498
Rework labor	120	120	122	391
Tote labor	1	1	2	3
Total lot cost	$287	$337	$851	$3,930
Finished units	805	918	956	978

related to the skill level required to perform each task. The bottom portion of Table 4.9 gives the amortized cost of the test systems and the total setup costs. Each test system is amortized over five years per working hour. The total setup is amortized over 12,000 PCBs. Table 4.10 is calculated from the data given in Tables 4.6 and 4.9. The costs per finished unit are (FBT) 36 cents, (ICT) 37 cents, (ICA) 89 cents, and (LBS) $4.02.

Employing the data of Tables 4.9 and 4.7, Table 4.11 is calculated. The amortization system cost for an in-circuit tester screened by a loaded-board shorts tester is essentially the same as the cost of an in-circuit tester alone. However, screening the in-circuit tester with an in-circuit analyzer reduces operational requirements of the in-circuit tester resulting in an approximately five percent reduction in cost.

The setup cost for the in-tandem configuration is the sum of the setup costs for each of the test systems in the combination. In both instances, the test labor cost is higher than that for an in-circuit tester alone. The rework cost is less for the LBS-ICT and 6 percent more for the ICA-ICT. The labor total cost is the same as employing an LBS-ICT and twice the amount for employing an ICA-ICT.

TABLE 4.11. Recurring Cost — In-Circuit.

SCREENED IN-CIRCUIT TESTER

	LBS	ICT	ICA	ICT
	REWORK	REWORK	REWORK	REWORK

Amortized cost		
• System	$ 281	$ 269
• Set up	490	510
Tester labor	93	106
Rework labor	120	130
Tote labor	3	6
Total lot cost	$ 987	$1,021
Finished units	956	963

TABLE 4.12. Recurring Cost — Functional.

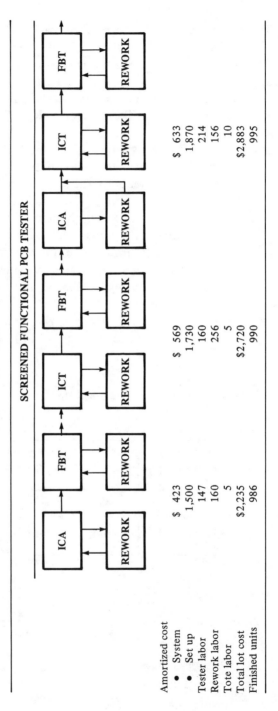

SCREENED FUNCTIONAL PCB TESTER

Amortized cost			
• System	$ 423	$ 569	$ 633
• Set up	1,500	1,730	1,870
Tester labor	147	160	214
Rework labor	160	256	156
Tote labor	5	5	10
Total lot cost	$2,235	$2,720	$2,883
Finished units	986	990	995

The cost per finished unit, as compared to an ICT alone, is 15 percent higher for an LBS-ICT and 19 percent higher for ICA-ICT. The recurring cost summary of Table 4.12 is calculated using the data of Tables 4.8 and 4.9. There is a clear reduction in the amortized system cost, the test labor, and the repair labor, along with a slight increase in the amortized setup cost and total labor. The results are a significant savings in finished unit cost, from $4.02 for a functional tester alone to $2.27, $2.75, and $2.89, respectively.

These calculations are based on 1000 controller boards with specific first-pass-yield and board volume. The functional board tester has the best performance in high first-pass-yield situations. As the first-pass-yield decreases, the effectiveness of the functional board tester declines rapidly due to the inherently longer diagnostic and rework time. Screening the functional board tester with an in-circuit analyzer results in a large decrease test and diagnostic time.

The effectiveness of an in-circuit tester increases dramatically as the first-pass-yield declines. When the first-pass-yield is 50 percent or less, an in-circuit tester produces the best all-around performance. The loaded-board shorts tester has an extremely low cost/performance ratio and fast test speeds. However, unless shorts are the dominant faults, it is not practical to employ a shorts tester in a high first-pass-yield situation.

5
THE REWORK STATION AND NETWORKING

This chapter discusses the evolution of the rework station from the classic, manual diagnostic/repair function to the computer-aided repair station. Topics covered include rework efficiency, effectiveness, and time. Also a discussion on low- and high-speed networking and a test area management system is included. The intent is to cover all the facets of a PCB production test facility.

5.1. THE REWORK STATION

If all the PCBs manufactured had 100 percent first-pass-yield, there would be no need for rework stations. Unfortunately, this is not the case. PCB rework is a necessary, time consuming, costly activity. The most important rework station consideration is the quality of repair. This consideration includes safeguards against introducing additional faults in the rework process, and documentation of the fault history to enable the manufacturer to correct the fault at the source. There must be accurate and reliable communication of the failure message from the tester to the rework personnel. As we shall discuss later, isolating the actual fault takes a certain amount of time and this time is affected by the skill and technical level of the operator.

The classic test/rework sequence follows a pattern. The tester's operator, upon identifying a failure, attaches a rework tag to the PCB and copies the failure message on the first line. With an in-circuit tester, the failure message would identify the shorted traces or the failed component. With a functional tester, the failure message would very likely indicate the failing node(s), which would give the rework operator a starting point in isolating the fault.

For faulty components identified by an in-circuit tester, the rework operator would read the rework tag and with the aid of production documentation would determine the location of the faulty component. Next, by flipping the board back and forth, part-side to solder-side, the operator determines which solder pad/leads need to be desoldered to remove the faulty component. The faulty component is desoldered and a replacement component is obtained. The replacement part is oriented and inserted into the PCB. The leads are soldered and trimmed, the rework tag is marked to indicate the fault correction, and the reworked board is cycled for retest.

For a short, the rework operator would again read the tag and, employing production documentation in conjunction with visual scanning and/or an ohmmeter, would isolate the shorted traces. After the short is desoldered and the corrective action noted on the rework tag, the reworked PCB would be returned to the in-circuit tester for retest.

The same procedure is followed with a functional tester, except that the failure message must be interpreted and benchtop instrumentation is often required for isolating the faulty component or digital device. Independently of the test system employed, finding and clearing a short is generally more difficult and time consuming than finding and replacing a defective component or device, because traces must be visually tracked and probed to locate a short between them. Fig. 5.1 illustrates typical rework time as a function of the tester employed. Note that the time to find and clear shorts and opens with different in-circuit testers is essentially the same, while the time with a functional tester can be significantly longer. This difference is due to the resolution of the failure message.

For the reworking of analog and digital components, the in-circuit tester requires the least amount of time while the functional tester requires the greatest amount of time. This difference is due to the resolution of the failure message.

Generally, rework efficiency is expressed in terms of the length of time in minutes required to repair a particular fault class divided by the rework yield expressed in the number of PCBs.

5.2. REWORK EFFECTIVENESS

Rework effectiveness is a measure of the ability to correctly identify and rework the faults on a PCB without introducing any additional faults. It is a function of the resolution and accuracy of the failure message, operator skill, knowledge and experience with the specific

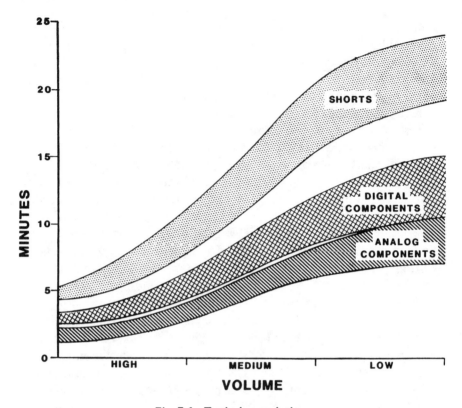

Fig. 5.1. Typical rework time.

PCB, the complexity of the PCB, the availability of rework aids and replacement parts, and the rework skill of the operator.

The average rework effectiveness determined by a Fairchild survey was 85–98 percent for high-volume production, 75–95 percent for medium-volume production, and 65–85 percent for low-volume production.   Fig. 5.2 illustrates typical rework effectiveness for the three relative production volumes versus the PCB tester employed.

The in-circuit tester provides the highest accuracy and resolution in the failure message, followed closely by the in-circuit analyzer, then the shorts tester, and finally the functional tester. A hypotheti-

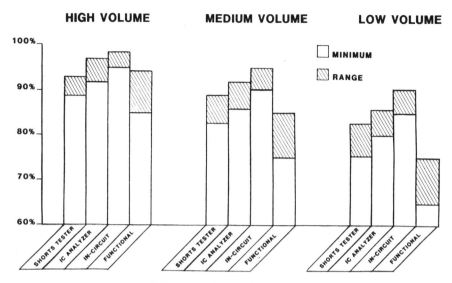

Fig. 5.2. Typical rework effectiveness.

cal example of rework efficiency is expressed in Fig. 5.3, illustrating 100 defective PCBs at an 80 percent rework effectiveness. When the original 100 PCBs were reworked and retested, 20 percent were returned to the rework station. When that 20 percent were reworked and retested, the same 80 percent effectiveness continued. Therefore, by taking the sum of the number of PCBs reworked in each iteration and dividing that sum by the initial number of defective boards, we obtain a value of 1.25 rework loops.

100 DEFECTIVE PCBs AT 80% REWORK EFFECTIVENESS

REWORK 100 PCBs
RETEST 100 PCBs
REWORK   20 PCBs
RETEST   20 PCBs
REWORK   4 PCBs $\dfrac{100 + 20 + 4 + 1}{100} = 1.25$ REWORK LOOPS
RETEST   4 PCBs
REWORK   1 PCB
RETEST   1 PCB

Fig. 5.3. Rework effectiveness.

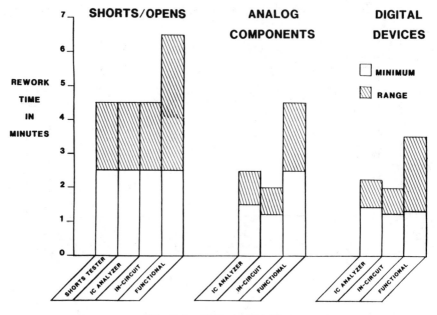

Fig. 5.4. ATE rework time.

Figure 5.4 illustrates the typical rework time as a function of PCB production volume.   These manufacturers employed a combination of commercial and in-house PCB test systems.   The curves illustrate that, at high volumes, the repair time required for shorts, digital components, and analog components is extremely low as compared to the time for lower volumes. Fundamentally, this results from the operator knowledge and experience with specific PCBs, and the resolution and accuracy of the failure message.

5.3. PAPERLESS REPAIR SYSTEM

Around 1976, one ATE company introduced the first paperless repair system in the form of a remote VDT in the tester's background. The test operator would enter the PCB type and serial number using a bar code reader wand. When the test system identified a failure, that failure was logged and the test operator was directed to place the defective board into the rework rack.   The rework operator would enter the PCB type and serial number via the remote VDT keyboard, and the VDT screen would display the PCB failure message, test and rework history, and, in the order of probability, the historic faults for the specific failure message. The rework operator would isolate and repair the fault and then enter the failure correction information.

This paperless repair system eliminated errors due to incorrect, incomplete, or illegible handwritten failure messages. In addition, this system supplied historic failure data to assist the operators in isolating the faults, and maintained a failure log for that particular PCB. The information was available to production engineers in a little less than real time to enable them to tune their manufacturing process to eliminate failures at the source rather than waiting for them to be found and repaired at the tester rework station.

Because of the cost/performance ratio and initial setup cost, the paperless repair system became popular only in high-volume PCB manufacturing companies. One high-volume production manufacturer stated that, with the paperless rework station, the standard rework times were 1.75 minutes for shorts, 1 minute for analog components, and 1.25 minutes for digital components. The accrued savings for the paperless repair system resulted in a payback period of less than one year.

5.4. COMPUTER-AIDED REPAIR

In 1981, an ATE company introduced the first computer-aided repair (CAR) system targeted for the medium- and low-volume production manufacturers. This CAR system consisted of a standalone computer system with mass storage, VDT, and color graphics display monitor. If the CAR system was not interfaced to the tester, the test operator would employ the classic rework card method of transmitting the failure message to the rework station where it was entered via a keypad. If the CAR system was interfaced to the tester, the rework station would receive the information via the paperless repair system.

At the rework station, the operator would either manually enter the PCB type and serial number or use a bar code reader. If the failure message indicated a short or open trace, the PCB outline would be displayed in green on the color graphics monitor; the PCB trace artwork would be displayed in blue and the appropriate traces would be highlighted, one trace appearing in red and the second trace in yellow. The contrast of different colors enables the rework operator to locate the fault traces more efficiently. If the failure message referred to a component, the color graphics display would show the PCB outlined in green and the component silhouetted in blue. The component in question would be highlighted in red. In addition, at the top of the screen, a description of the failed component and any relative details would appear.

These types of displays aid rework operators in isolating faults. With a single keystroke, the operator may switch the PCB image from the solder side to the component side. The color combinations are chosen for the greatest contrast and can be used to overcome sight problems such as some forms of color blindness.

After isolating and repairing the fault, the rework operator would enter the fault correction information. With an in-circuit tester, the rework operator would call up the next fault on the PCB with a single keystroke.

Some rework strategies would not attempt to repair a PCB that contains an excessive number of faults. The actual number is based on the total cost of the PCB versus the cost to rework. In this case, the total number of faults contained in a PCB should be displayed first so that a decision can be made to abort rework or not.

The CAR system PCB program is generated on an XY-grid using interactive software and a crosshair cursor on a digital table. Alternately, the PCB program can be generated by a CAD system. Component geometry is selected from a parts library and positioned by the crosshair cursor. Component identification is entered via the keyboard along with any relative information.

Because of the long and costly preparation time required for generating the PCB artwork display, a modified CAR system entered the market. It used an outline of the PCB with positioned silhouettes of various components. For shorts, the components connected to the shorted traces would be shown in red and the specific lead would be shown in a flashing yellow. The general routing of the shorted traces are displayed in yellow. For a defective component, the specific device would be shown in flashing red. With this type of display, even a PCB with 250 components could be fully programmed within a day. The less sophisticated display helped clear the cost justification hurdle.

In early 1983, a second generation CAR system was introduced. This system enhanced the earlier CAR system by incorporating a semi-automatic production assembly station. The semi-automatic assembly station portion consisted of a dual-beam projection system, and an automatic parts binning and delivery subsystem. The dual-beam projection system is used to automatically locate components and component pads/leads to be desoldered. It also employed a blinking beam to indicate polarity. For an IC, the system would highlight the two extreme and opposite leads. This would allow the operator to concentrate on the PCB and not be disoriented by looking at a graphic display terminal. When the board is flipped from the component side to the solder side, the dual-beam projection system would follow.

The automatic parts binning and delivery subsystem would automatically obtain the required replacement parts from a series of bins and deliver the part to a replacement part dispenser. This minimizes the possibility of inserting the wrong part. Further, the color display was modified to allow zooming in on the faulty area. The outline of the faulty component is positioned in the center of the viewing area. When traces are shorted together, the faulty information would determine what net was shorted and the display would show those traces.

The vendor improved upon the preparation software by offering a programming service for the color graphics display and allowing the dual-beam projection system to be independently programmed by the use of XY-coordinates.

The vendor estimates a reduction in rework time of 50–60 percent when this generation of rework station is employed.

5.5 NETWORKING

Networking is a two-way data link between a computer and another computer or video display terminals, a programming station and three or more programming consoles, a programming station/host computer, and testers and rework stations, a main tester and a satellite tester or quality control station/tester and a test management system, a test area supervisor or a test area management system.

Networking is generally classified into two categories: low speed or batch transfer, and high speed or real time. The low-speed type usually employs a RS232 interface at a rate of 300 baud to 19.5 kilobaud per second. The high-speed type may employ an Ethernet or cabling system at 10 MHz rate. (Ethernet is a product of Xerox Corporation, Stamford, CT.)

5.5.1. Low-Speed Networking

The initial concept of networking expanded the capability of a programming station by three to four times shared programming consoles. This concept could be expanded further to production by having the programming station located in the test engineering department downloading test programs to the testers on the manufacturing floor, thereby controlling and managing the test program's integrity. This strategy requires some discipline, since most of the engineering changes are often performed on the tester and not necessarily incorporated into the master program in the programming station. This minor problem could become a serious pitfall. Several companies, in order to avoid this pitfall, have the new test program automatically uploaded to the programming station and a warning flag set. A future fetch command will result in a message to contact test engineering.

Network strategy has expanded further to include downloading test programs to remote facilities at a 300–9600 baud rate via telephone modems.

The quality control department became interested in capturing failure data for immediate action, so users started uploading failure data information to the programming station for data analysis and management report generation. This uploading required additional software plus large mass storage units. The programming station started acting like a host. The limiting factor is the speed of data transmission. Downloading a test program and retrieving failure data once or twice a day at an individual batch transfer time of 4–20 minutes for 1–4 testers is acceptable to most production managers.

Another common use of the low-speed networking is to provide a data link between the main tester and satellite testers. The main tester is a full-blown system where test programs are generated, debugged, and then stored. The satellite testers are minimum-point-count systems with little, if any, mass storage capability. The strategy is to cut the capital costs by having one test system that can test all the products and smaller testers that will each test a section of the product line. This is particularly useful for high-volume, low-pin-count PCBs. The test program is downloaded to each satellite for the specific PCB lot, and the failure data are uploaded to the main tester for data analysis and report management via high speed network link.

5.5.2. High-Speed Networking

High-speed networking (Fig. 5.5) allows a rapid transfer of data. This transfer is virtually transparent to the user. The standard Ethernet coaxial cable comes in segments with a maximum length of 1640 feet, which may be expanded using repeaters to any number of cable segments. Ethernet specifications allow up to 100 network interfaces per cable segment. The distribution of network intelligence and control is via a microprocessor in each network interface. The transceiver controls the flow of data on and off the cable employing a collision detection algorithm. This is faster and more reliable than transporting disk packs to and from programming stations, testers, and/or host computers. The high-speed network interfaces to any RS232 serial port or any 8-bit parallel port, thereby providing flexible linkage for a variety of equipment and manufacturers.

In addition, most networks employ cyclic redundancy checks (CRC), blocking corrective polynomial circuits for detecting and recovering garbage transmission. The high-speed network may be utilized in any of the low-speed networking applications plus receiving and sending data to quality control stations located throughout the production line. This enables the test manager to monitor and control production, thereby providing the ability to remove a problem at the source rather than identify it at the board test station.

5.6. TEST AREA MANAGEMENT SYSTEM

A test area management system is a management tool for automatically collecting manufacturing data to improve management's view of

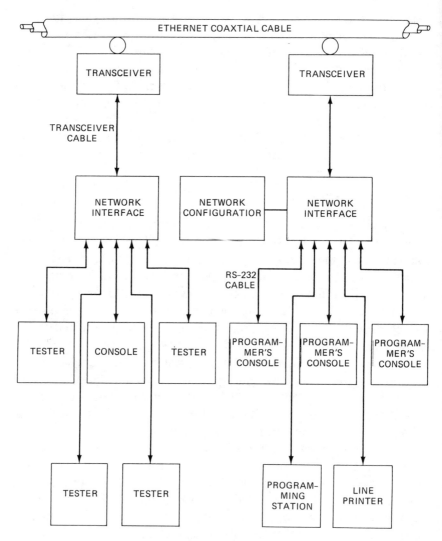

Fig. 5.5. High-speed networking.

the manufacturing process, and to provide accurate real-time data for manufacturing decision-making. The data collected is primarily from testers, rework stations and terminals. Upon collection, this data can provide the basis for accurate and timely reports along with real-time and trend alarms that will expose production problems. The data will provide management with a tool for improving controls of their production operation, the manufacturing process, flow of goods, the integration of automatic test systems, and selection and development of manufacturing information systems. The implementation of an in-circuit tester in the manufacturing board test has proven to be an effective means of reducing production costs. With the reduction of direct labor costs, the question arises, how do we reduce the larger indirect costs associated with the production process? Many believe the answer is a test area management system that provides management with immediate indications of the problems as they occur.

The test area management system provides automatic generation of rework instructions, automatic logging of test data and rework activity, downloading of test programs, remote data transmission, computer-to-computer communication links, automatic program generation, and automatic generation of the maintenance history. All PCBs are assembled, soldered, tested, reworked, and logged into work in process or finished goods inventory. If every production operation were controlled automatically, production problems would be identified as they occurred and the solution could be implemented immediately.

Certainly not all operations can be performed automatically without human intervention. Receiving inspection assembly and inventory control consist primarily of manual operations requiring interpretation and decision-making. Further, the data obtained in these operations is used for the production control functions and can easily be entered, by video display terminals, into the test area management system. During production testing, most automatic test equipment can capture test data. The test area management system takes this data and produces manufacturing reports such as production yield, test time, rework time, test-station throughput, rework-station throughput, inventory levels, work in process, different component failure histories, and task time standard − all reported on a demand basis. Cost savings can be achieved by improvement in the information sent back to the manufacturing floor. Reductions in indirect costs can be achieved by production and the effort required by management to track the flow of boards through the production process, and in the effort required to define production problems.

The system's hardware consists of standard building blocks that may be interconnected as required to monitor and control the manufacturing process. These building blocks are essential disk storage media, video display terminal, high-speed printers, computer-to-computer interface, magnetic tape storage media, and barcode readers.

5.6.1. System Software

The most important feature of a test area management system is the effectiveness of the software. The software operating system should be user oriented in high-level English-language automatic program generation for user's test programming. The acquisition of test data is performed automatically. The central historical file is managed in the system's mass storage unit and the data are based upon all control activities. The application program accesses data from the historical files to generate rework instructions and maintenance status reports. Each company has its own concept of what a test area management system is, what data it should automatically store, and how the data should be reported. Hopefully the difference is in software. Let us discuss a test area management concept.

The test area management system software includes a data base manager, real-time process control, test program control, and report generation.

5.6.1.1. Data Base Manager. The data base may reside on a number of storage media. The data base size varies with user's requirements. The data base manager consists of real-time data collection, automatic data storage, retrieval and security.

Real-time data collection allows real-time control over the manufacturing process and production testing. The data base manager stores all current PCB data. Once stored the data may be retrieved and presented in raw form or recombined to provide a specific report.

The test area manager's host recognizes when the data base is near capacity without interrupting the network. A capacity alarm, set at 80 percent sends a warning flag to the test area management console to purge unnecessary data. At 90 percent capacity, another alarm is sent to the test area management console and to all network terminals. At 100 percent data base capacity, an alarm is sent throughout the network and the data storage is routed to a tester's mass storage unit.

The data base monitor is a mechanism to automatically store inactive data on off-line media. This automatic off-line storage utility prompts the operator for data age before it is stored off-line. Inactive PCBs that are idle because of a parts shortage are excluded from the aging process. After the off-line storage transfer is completed, the utility will inquire if there are any data backup requirements. If yes, the data are stored on the same type of medium. The off-line data and/or backup data is able to reenter the current active data base in whole, or in part, by the appropriate retrieval commands.

The cumulative data base is difficult to define and careful consideration must be given to its implementation. The cumulative data base is a rolling time function – one week, one month, one quarter, or one year. Overall, the cumulative data base is to obtain detailed current information, normalized by history for various trend reports: process yield, test station yield, rework efficiency, defect analysis, and production throughput.

Without security, changes could be made to the data base or other files without management's knowledge or permission. To minimize access to the data base manager, every user must have identification and a user priority assignment number. The proper identification gives the operator access to the data base, test programs, report generation, and tester operation.

5.6.1.2. Real-Time Process Control. The real-time process control tracks every PCB throughout the production process. If a PCB is inadvertently routed to the wrong location when the PCB is logged, the real-time process control will redirect the PCB to the proper location. Further, the real-time process control monitors the number of rework cycles and defects encountered. When the PCB exceeds a user-determined limit, a warning alarm is sent. Real-time alarms provide immediate control over the manufacturing process and test operation. Generally three yield alarms and two trend alarms are established. A yield alarm is set for each test station to ensure proper tester operation. Second, a yield alarm is set at each test station for a specific PCB to monitor the test program. Third, a yield alarm is set at each test station to monitor reworked PCBs efficiency. One trend alarm tracks consecutive PCB failures within a lot, monitoring the manufacturing process. The other trend alarm tracks the transition time between data entry stations monitoring the process flow.

All alarms prompt the system manager to a specific alarm report that is both displayed on the appropriate terminals and automatically printed out on the appropriate printers. The alarms are also displayed at the affected tester(s).

5.6.1.3. Information Processing. After the data are collected and stored, they must be reformatted and reused to benefit production. The information processed includes test program revisions, test station, and PCB information. The test program revision employed during PCB testing is tracked to ensure proper quality of testing. Every operator needs to log onto the test area management system using an individual identification number. The operator identification is checked against a master log. Only one operator can be logged at a production station at a given time. The resulting data obtained from that production station tracks both the operator's efficiency and the station utilization.

The individual board identification is entered by a barcode reader prior to assembly, soldering, test, rework, and retest. Listed under the specific board serial number is all the production activity and history: when and where it was tested, the failures incurred, and the rework station repair activity.

Paperless repair and rework are two independent functions. The repair function displays on a maintenance terminal a tester failure, maintenance history, and probable faults. The rework function is the failure data of a specific PCB displayed on a rework station terminal. In both cases the operator provides reentry into the respective data base, the actual failure correction information.

5.6.1.4. Test Program Control. File transfer capability is provided from the host to any tester and from tester to tester. However, a password is required or the data base manager will inhibit access. File transfer accommodates both ASCII and binary files.

The test area management system provides the user with the option of storing both ASCII and binary files. File storage is optional by default because data base requirements supersede file storage. Under normal conditions the host is the master test program storage medium.

At the host there are main and secondary test program directories. The main directory contains a listing of different PCB test programs for the different board testers. The secondary directory contains test program revisions, with the programmer's name and revision dates. The operator selects the test program for a particular PCB and tester plus the revision level. The default is the latest test program revision.

5.6.1.5. Report Generation. The report generation facility provides access to company standardized reports and the tools to generate report formats. Some access identification limits the operator to a group of company standardized reports. The operator is prompted by menus asking the appropriate question to produce the specific report desired. Further, the operator may delete old reports or append new reports after the old report in the same file. The printout is properly paginated and spaced for management review. Standardized reports provide the operator with required daily information. A utility of the report generator is on-demand reports. On-demand reports are required only when certain conditions exist. The operator is queried as to the reporting medium (terminal or printer) and authorization numbers. This gives the operator access to all previous reports and files, both in active and off-line storage.

Another utility of the report generator is automatic reports. The automatic reporting facility provides cyclic reports at specific times and locations, and in specific reporting media.

The format of the reports is a function of the parameter set in the user's design application program. Generation of these reports is simplified by the use of the system's programming aids and the automatic program generator.

An example of the production yield report for the PCB test station is demonstrated in Fig. 5.6.

Two PCB manufacturing and tests are exhibited. The first section of the product yield report contains the part number 781674, a 74R

PART NUMBER	781674	781662
PART NAME	74R Processor	VZ Interface
# LOT NUMBER	11424	12413
MFG LINE	01	03
IN-CIRCUIT TESTER	04	01
REPAIR STATION	03	02
LOT SIZE	112	179
SET UP TIME (HR)	4.21	3.55
ASSY & SOLDER (MIN)	282.2	140.3
FIRST-PASS-YIELD	56.3%	63.1%
SECOND-PASS-YIELD	33.0%	21.2%
THIRD-PASS-YIELD	7.1%	15.6%
FOURTH-PASS-YIELD	3.6%	0.0%
TOTAL TEST TIME (MIN)	104.9	91.2
AVERAGE TEST TIME (MIN/PCB)	0.94	0.51
DEFECTIVE PCBs	49	66
PCB REWORKED	65	94
FAULTS/PCB		
TOTAL REWORK TIME (MIN)	168.6	105.4
AVERAGE REWORK TIME (MIN/PCB)	3.44	3.11
TOTE TIME (MIN)	48.6	28.4
PRODUCTION TIME (HR)	10.7	7.76
THROUGH PUT (UNITS/HR)	11.12	23.08

Fig. 5.6. Production yield report — PCB test.

processor, lot number, production line number and the in-circuit tester employed along with the repair station. The second section covers the lot size of 112 PCBs to be manufactured. The production line setup time is 4.21 hours. Assembly and soldering time is 282.2 minutes. The results of the manufacturing process were a first-pass-yield of 63 boards, or 56.3 percent. After reworking the 49 defective boards, 37 boards passed the second test sequence for a yield of an additional 33 percent. The remaining 12 boards were returned to rework. The third retest resulted in eight boards passing. After rework, the remaining four boards passed, completing the production yield of 100 percent. The total test time was 104.9 minutes for an average test time of 0.94 minutes per PCB. Forty-nine boards were tested bad; however, due to multiple faults the rework station logged 65 boards, of which 12 repeated once and four repeated twice. The total rework time was 168.6 minutes for an average rework time, based on 49 boards, of 3.44 minutes. The tote time is logged at 48.6 minutes, producing a total production time of 10.7 hours, resulting in a throughput of 11.12 units per hour.

Part number 781662, a VZ interface, is also exhibited, with a lot size of 179 units and a first-pass-yield of 113 units or 63.1 percent, a second-pass-yield of 32 units or 21.2 percent, and a third-pass-yield of 28 units or 15.6 percent. The total time was 91.2 minutes, resulting in an average test time of 0.51 minutes per printed circuit board. The rework time of 205.4 minutes resulted in an average rework time of 3.11 minutes per PCB. The total production time was 7.76 hours, or 23.06 PCBs per hour.

6
IN-CIRCUIT TESTING PHILOSOPHY

As previously stated in Chapters 2 and 3, the in-circuit tester is a manufacturing verification tool. It tests individual components and their interconnections on a printed circuit board. The basic assumption of this strategy is, if the board design is proven and the artwork and components are correct, the board will be functional.

The in-circuit tester employs a guarding principle to measure the performance of individual components by electrically isolating the UUT from the surrounding circuitry by means of a bed-of-nails fixutre. The in-circuit tester will identify multiple faults within each of its four or five levels of test hierarchy: shorts, passive components, IC orientation, digital logic, with a few testers testing active analog devices. At each test level all the testing is performed and faults stored before a failure flag is set to halt the test and display the failure message. The first two levels of the test hierarchy, shorts and passive components, are performed on an unpowered printed circuit board. Identifying and removing all the shorts and defective components before power is applied to the PCB removes the hazard of damaging the PCB and blowing good parts. In addition to the safety feature, it is extremely difficult to measure individual components when power supply current is flowing. The remaining test hierarchy levels are with the circuit board under power. The in-circuit test hierarchy is shown in Fig. 6.1. The test times per level are a function of the in-circuit tester employed and the number of components per PCB.

6.1. UUT/FIXTURE VERIFICATION

UUT/fixture verification is to ensure all of the test fixture probes are making solid electrical contact with the UUT. One source of unverifiable faults is caused by test probes not making contact with the UUT. Picture a bent test probe that rotates a few degrees each time

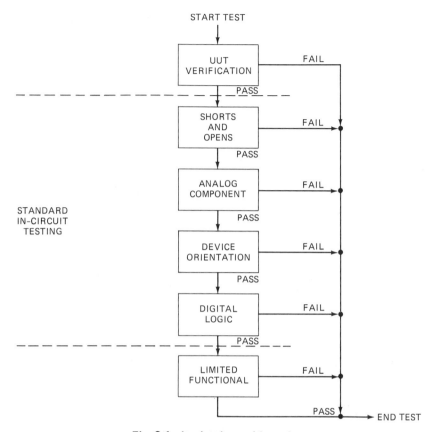

Fig. 6.1. In-circuit test hierarchy.

the test fixture is actuated. The result is random missed contacts. The random opens cause good parts to be identified as failures. Another example is a test fixture's tooling pins becoming worn, causing the PCB registration to be offset. This would result in one or more test probes not making proper UUT contact. In-circuit testers that have UUT/fixture verification capability generally perform the test before executing the system test hierarchy, which ranges from 10 to 20 microseconds per test probe. UUT/fixture verification is a gross or catastrophic measurement between each fixture's test probe and the representative UUT node. Problems such as probe contamination, surface flux on the PCB, and bad registration can all be determined prior to performing the actual PCB test. Buildup of contamination on the test probe tip will not be detected until the resistance is in the high kilohm range.

The UUT/fixture verification test electrically floats the UUT by applying the same voltage to both the ground and power buses. Then each UUT node is scanned for a voltage. Each of the test points in contact with a UUT node will be the center point of a voltage divider, which is determined by the resistance path on the UUT either to the power or ground bus and the measurement amplifier's input resistance. The measurement voltage threshold is typically set at a 10-to-1 ratio. That is, if the applied voltage is 5 volts and the input impedance of the measurement amplifier is 1 megohm, for a 10-megohm UUT resistance path the measured voltage would be 0.5 volt, below which a defective contact is flagged.

In the interest of reducing good-board test time in a medium first-pass-yield situation, some test engineers will only employ the UUT/ verification routine when a failure is identified to ensure that it is a true failure, not a test probe contact problem. This is a good test strategy for a low number of average faults per PCB.

6.2. SHORTS/OPENS TEST

The first level of testing is detecting shorts. The test time ranges from 20 microseconds to 15 milliseconds per test point depending on the type of switching matrix and shorts detection algorithm employed. Shorts detection is accomplished by one of two methods. One is digital short detection, where the sensor has a low voltage threshold at a given driver current which determines the resistance crossover point. This method is fast but threshold limited. The second is analog shorts detection, which actually measures the resistance value and performs a resistance crossover comparison. This method yields accuracy at the price of test time. The basic shorts algorithm is to apply a voltage of 200 millivolts to node 1 and measure the voltage on nodes 2 through N at an equivalent threshold of 10–200 ohms. If the voltage measured does not exceed the crossover threshold, node 1 is open. Then the shorts algorithm applies the voltage to node 2 and measures the voltage on nodes 3 through N. When a voltage is measured that exceeds the crossover threshold, those two pins are shorted.

A faster shorts detection algorithm is to apply the voltage to all the nodes except for node 1 and then measure node 1 for a voltage. If no voltage is measured, node 1 is open. If a voltage is measured, node 1 is shorted and node 1 is flagged. The voltage is then removed from node 2 and applied to node 1 if flagged as a short. If the node was open, it remains open. Node 2 is measured for a voltage. If none is measured, node 2 is flagged open. The voltage is then removed from node 3. Node 3 is measured for a voltage. If none appears, the voltage is removed from node 4, and node 4 is measured for a voltage, etc., through N nodes. At Nth node the algorithm goes into the second phase, which measures each flagged shorted node in sequence as in the basic shorts algorithm to identify the specific shorted nodes.

A third shorts detection algorithm is to select a block of nodes (say, 1 through 50) and another block of nodes (say, 51 through 100), apply the voltage to the first block, and measure for a voltage on the second block. If no voltage is measured, each block is divided in half (nodes 1 through 25, 26 through 50, 51 through 75, and 76 through 100). Voltage is applied to three of the blocks and a voltage is measured on the fourth block. If no voltage occurs, the blocks are then subdivided in half again, and again. When a short is detected, that specific block is divided in half and a voltage is applied to one half of the block and the voltage measured on the other half of the block, etc., until the shorted nodes are detected. A verification is then performed as in the basic shorts algorithm.

Typically the allowable crossover range for loaded-board shorts detection is between 10 and 200 ohms. The default value for no crossover specified is 10 ohm. The low value of resistance is to ensure it is a short, not an equivalent low-value parallel network. The allowable source voltage ranges between 50 millivolts and 1 volt. The default value for no source voltage specified is typically 200 millivolts. The low applied voltage is to ensure that none of the semi-junctions will be forward biased, which may cause erroneous shorts being detected.

A subdivision of shorts detection is continuity testing. In-circuit continuity testing on a loaded printed circuit board may be trace continuity testing or open testing for jumpers and fuses. Trace continuity testing requires a fixture test probe at every extremity of each trace to ensure the integrity of each trace. The additional number of test probes increases the total test time and may add unwanted capacitance. However, the test comprehensiveness is increased and diagnostics are more specific. The tester will identify an open trace rather than a bad component or series of bad components. However, because of the extremely low percentage of open trace faults and large increases in fixture cost, and the possibility of capacitance loading effects, in-circuit testing predominantly employs one fixture test probe per trace and performs continuity tests only for jumpers and fuses. The continuity test methodology is identical to the shorts test except the threshold levels or crossover resistance value may be set in the 10 ohm to 1 megohm range depending on the application. The test time is faster than for shorts testing, ranging from 10 microseconds to 10 milliseconds per test probe.

The next test hierarchy level is individual passive component measuring and/or testing. The measurement and/or test time ranges from 100 microseconds to 3 seconds per component depending on the type of switching matrix, the test algorithm, and the circuitry on the unit-under-test. Passive component detection is separated into two categories, measure and test. Measuring a component is generally accomplished by applying either operational amplifier or impedance bridge technology, where the actual value of a component is determined and compared to a stored set of limits. Testing a component is accomplished by calculating the component's value at a specific time or contained in an equivalent circuit, then measuring the component under these conditions and comparing the value to the stored calculated set of limits. Testing a component is a trade-off of accuracy for speed which will be discussed in more detail later in this chapter.

6.3. ANALOG MEASUREMENT

In an operational amplifier mode, measuring two terminal components out of circuit is typically accomplished by either applying a known voltage across the component to be measured and measuring the current produced, or applying a known current through the component to be measured and measuring the voltage produced. When measuring low value components where the lead impedance is a concern, Kelvin connections are employed to sense the signal at the component

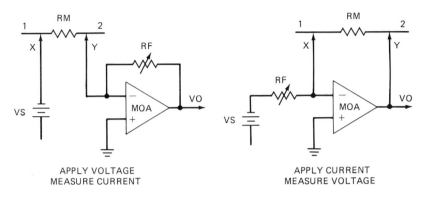

APPLY VOLTAGE   APPLY CURRENT
MEASURE CURRENT   MEASURE VOLTAGE

Fig. 6.2. Two-terminal measurement.

terminals and provide a correction value to offset lead errors. Fig. 6.2(a) shows a known voltage being applied. The resulting current through the resistor to be measured (RM) is then determined by using a measurement operational amplifier (MOA) as a current-to-voltage converter. The measuring operational amplifier is a high-gain difference amplifier with two high-impedance inputs and a low-impedance output. Applying a signal to a minus input with the plus input reference to ground causes current to flow through the reference resistor RF in the feedback loop to cancel the current through the minus terminal. Current flows through the feedback loop until the voltage potential at the minus input returns to virtual ground. Thus RM equals VS divided by VO multiplied by RF. Fig. 6.2(b) illustrates the known voltage source VS divided by the reference resistor RF. This combination results in the fixed current source being applied to the device-under-test (RM) in the MOA feedback path. The MOA output voltage VO is proportional to the DUT value. RM equals VO divided by VS, multiplied by RF. Using an AC voltage source adapts the testing technique to capacitors and inductors. The major difference between the operational amplifier and impedance bridge technology is measurement speed. Employing an operational amplifier is generally 20–50 times faster.

6.3.1. Three-Wire Measurement

In measuring individual components in-circuit, the major concern is isolation of the device-under-test (DUT), from the surrounding

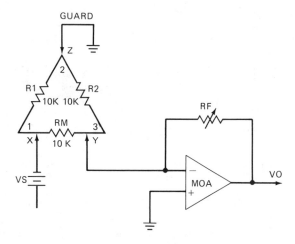

Fig. 6.3. Guarding principle.

circuitry. The method employed is the principle of nulling currents, referred to as guarding, illustrated in Fig. 6.3. The diagram shows the resistor-under-test (RM), a 10K resistor, in parallel with R1 and R2, both 10K resistors. RM would measure the network of 20K in parallel with 10K or 6.6K. However, by placing the guard, (ground) at point 2, the current through R1 is 0 as points 1 and 3 are essentially at ground potential. R2 provides negligible load because of the nearly zero-ohm output impedance of the MOA. The output voltage is proportional to the RM/RF, programmed to be 10K. The accuracy of the three-terminal measurement is typically 1 percent from 100 ohms to 1 megohm and drops off rapidly beyond both ends of the range. The inaccuracy at the low-range values is due to lack of Kelvin measurement. The inaccuracy at the high-range values is due to the inability to stop current from flowing in paths with significantly smaller impedance than that of the component to be measured.

The amount of current flowing in the guard path versus the amount of current flowing in the measurement path is referred to as the guard ratio. For a three-terminal measurement, the guard ratio is typically between 100 and 50 to 1. If a 500-ohm resistance is guarded, the maximum resistance that can be measured without degradation in accuracy is between 50 and 25 kilohm. The main advantages of three-wire measurements are test speed, ease of fixture fabrication, and system price.

6.3.2. Four-Wire Measurement

There are two methods of four-wire measurement. The first method is a two-wire Kelvin connection, force plus sense and measure plus sense. The other method, the more common in-circuit technique, is force, measure, and ground plus sense. This method may be thought of as Kelvin ground connection which eliminates the error introduced by the matrix resistance in the guard lead. The ground sense addition ensures a significant reduction in parallel path current and minimizes the error current, providing a more accurate high-value measurement capability. Typically the 1-percent high-value range is extended to 10 megohms. Typically for four-terminal measurements the guard ratio is between 500 and 1000 to 1. If a 500-ohm resistor is guarded, the maximum resistance that can be measured without degradation in accuracy is between 250 and 500 kilohms. However, if two or more guard points are used to isolate a component, individual guard current loops will be established, producing more parallel current paths resulting in a measurement error difficult to calculate.

6.3.3. Five-Wire Measurement

The five-terminal measurement matrix, consisting of a force plus sense, measure plus sense, and guard, is not generally incorporated in commercial in-circuit testers as the guard ratio is the same as in a three-terminal system at a higher cost and complexity. The advantage of five-terminal measurement is the ability to provide a guarded Kelvin or near-Kelvin measurement capability which would expand the low measure range to less than 10 ohms and/or the ability to measure precision resistors.

6.3.4. Six-Wire Measurement

The six-terminal measurement matrix consists of a force plus sense, measure plus sense, and guard plus sense. It is employed for high-accuracy measurements, and increased measurement range from less than 1 ohm to 100 megohms. Typically the guard ratio is 10000 to 1; for example, if a 500-ohm resistor is guarded, a 5-megohm resistor could be measured without degradation in accuracy. The disadvantage of six-wire measurement is the slow test speed and test fixture flexibility.

6.3.5. Large-Capacitor Measurement

The most dynamic passive component measurement is capacitance. Generally the AC source is selected from 100 hertz to 100 kilohertz as the size of the capacitor reduces from 1000 microfarads down to 1 pico-farad. When measuring a high-value capacitor, the AC-to-DC conversion time produces an extension to normal test time. A measurement delay, referred to as wait time, is programmed in fixed time increments into the test statement to allow sufficient low-frequency samples to produce a stable measurement. The time required for the capacitor to become fully charged may be a few milliseconds to a number of seconds. If a capacitor takes 650 milliseconds to become fully charged, and the test algorithm wait time is in 500 millisecond increments, then the capacitor's test time is one second.

One software routine employed to reduce test time is automatic wait time, AWT. This routine samples the voltage response until the signal is stable before a measurement is made. Generally the maximum programmable automatic wait time is 32 seconds with a default condition of 5 seconds. If a stable signal is not detected in the allotted time period, an over-range measurement indication results and the test fails. Automatic wait time decreases the test time from the fixed-wait-time increment method, with no degradation to accuracy or resolution. If a capacitor is fully charged in 650 milliseconds, the capacitor's test time is 650 milliseconds.

6.3.6. Parallel Components

Parallel networks of like components are impossible to isolate for individual measurement using normal instrumentation. An accumulative value is the best the in-circuit tester is capable of performing. However, resolving parallel networks of unlike components may be resolved by multiple testing at various frequencies and calculating the individual values. A more efficient methodology is measuring the quadrature vectors of the current through and the voltage across the parallel network, then calculating the real and imaginary values for the measurement results.

6.4. ANALOG TESTING

Testing a capacitor by calculation is another method employed to reduce test time. The two most popular methods are fixed-time sample and impedance signature. The fixed-time sample algorithm has a maximum time-out test time of typically 30 milliseconds. To test a capacitor with a large RC time constant, say 2 seconds, the system will calculate the value of the capacitor at the 30 millisecond time-out point. During testing a voltage sample is taken at this point and compared to the stored calculated value. The accuracy of this test methodology varies with the instrumentation employed and the size of the capacitor. At the sharp rising portion of the capacitor charge response curve, the accuracy may vary from 20 to 40 percent with a resolution of about 10 percent. As the sampling point approaches the more stable portion of the capacitor voltage characteristic curve, the accuracy improves from 5 to 15 percent with a 5-percent resolution.

When testing the capacitor employing impedance signature, the system calculates an appropriate wait measurement time at which three voltage samples will be taken at 100 microsecond intervals on the capacitance charging voltage response curve. During test execution the system will measure the three voltage samples and calculate the value of capacitance. The three-voltage sample provides a better definition of the capacitor's voltage response curve as the capacitor is charging. If auto-learn is available, the system stores the wait measurement time and learns the three voltage samples. During testing, the system employs the same criteria by comparing the voltage samples to the stored values. For calculated capacitance, the typical test accuracy ranges from 5 to 10 percent with a resolution of 1 to 5 percent. For voltage sample comparison the typical test accuracy ranges from 5 to 15 percent with the same resolution at a significant reduction in test time.

6.5. OPERATIONAL AMPLIFIERS

Most in-circuit testers do not have linear device library models. Therefore, the test program must be generated manually. There are differences of opinion as to what constitutes a valid, in-circuit test for operational amplifiers. The output of the operational amplifier is determined by inputs and surrounding external circuitry. Typically, the test consists of three basic static tests: DC offset, DC gain and differential voltage.

6.6. ANALOG TESTING SUMMARY

Table 6.1 summarizes the most common analog measurements performed by in-circuit testers. The lowest-value elements are measured first in ascending order. In addition, when testing a passive component where the measurement nominal is near the high end of the range

TABLE 6.1. Analog Measurements.

Components	Tests
Shorts and opens	Crossover value
Capacitor (nonpolarized)	Nominal value ± tolerance
Electrolytic capacitors	Nominal value ± tolerance
	Leakage current (current orientation)
Diode – all types	Reverse leakage current
Diode – signal, LED, Zener	Forward voltage
Diode – Zener	Reverse Zener voltage
Field-effect transistors	Gate–source leakage current
	Drain–source leakage current
Inductors	Nominal value ± tolerance
	Winding resistance (over 1 ohm)
Integrated circuits	Orientation
Operational amplifiers	Closed-loop gain
	Closed-loop saturation
Relays	Contact closure – de-energized
	Contact closure – energized
	Coil and contact resistance (over 1 ohm)
Resistors	Nominal value ± tolerance
Silicon controller rectifiers	Forward (on) voltage
	Forward (off) leakage current
	Reverse voltage leakage current
Transformers	Winding resistance (over 1 ohm)
	Continuity
Transistors (bipolar)	Forward junction voltage (CB and EB)
	Reverse junction leakage current (CB and EB)
	ICES (collector leakage with emitter base shorted)
	DC gain (HFE minimum)
Unijunction Transistors	Interbase resistance
	Emitter reverse current

and the tolerance of the device would cause the upper limit to be in the next higher range, it is necessary to adjust the nominal value and the tolerance of the test. The procedure is to use the low end of the next highest range as the nominal and calculate the tolerance for the correct upper and lower limits based on the new nominal.

6.7. DEVICE ORIENTATION

Orientation tests for discrete components form a subset of the passive component test sequence either by measuring the components characteristics or performing a leakage test. In one or two in-circuit testers, IC orientation is performed on an unpowered board by mapping the IC's internal diode characteristics between the power and ground pin to the other pins. The benefit of this IC orientation technique is that the IC orientation is determined before power is applied to the UUT, thereby ensuring that the IC will not be damaged by a misapplication of power to inappropriate IC pins.

The most popular methodology is to have a designated IC orientation level in the test hierarchy. IC orientation is generally performed by pulsing each pin of the IC high and low. Revised ICs typically cause the input/output circuitry to now be forward biased to power and ground, and therefore would detect improper orientation of the device. The benefit of this IC orientation technique is speed. The total test time ranges from 100 microseconds to 500 microseconds per device.

6.7.1. Bus Testing

Some in-circuit tester's test hierarchy include logic bus verification as a subsection of digital IC orientation test. The logic bus test routine is to isolate bus faults caused by an IC device whose bused inputs or outputs are stuck-at-high or stuck-at-low state. If the individual ICs on a bus were tested and the bus was at a stuck-at-state, many, or all, the individual ICs would fail. For a tri-state bus all the parallel ICs outputs connected to the bus are disabled by applying the appropriate input vectors. With a pull-up resistor connected to the bus, the bus is tested for a logic high, then with a pull-down resistor connected to the bus, the bus is tested for a logic low. If the bus test passes, it is free of stuck-at faults. If the bus test is unsuccessful, failed logic-high or logic-low voltage is measured and stored. Then individually, each IC output is forced by the appropriate input vectors into the bus logic failure state and the voltage is measured and compared to the stored disabled voltage value. Any IC bus output causing little or no difference in voltage is flagged as the defective IC. If the bus test routine does not identify a failure at an IC output, the tester's software branches to a guided current probe subroutine. The VDT will display a message that prompts the operator to probe a specific IC input connected to the bus. After measuring and storing the current on that IC leg, the VDT will prompt the operator to probe another IC leg. This procedure continues until all the IC input leg currents have been measured and stored. Then a comparison is made of all the stored current values and the IC whose input draws the most current is flagged as the defective IC device.

Another bus test routine is to apply another subroutine that may have been called up before the guided current probe subroutine is IC pin continuity verified. Most in-circuit test fixtures are built with one test probe per trace. Therefore, the trace may be open between the fixture test probe and the IC leg. The VDT displays a message to prompt the operator to probe a leg of a specific IC. The probe grounds each IC leg and the system measures for a ground at the respective test point. No continuity identifies an open PCB trace IC pin connection.

Another bus test routine is to apply the appropriate IC inputs to disable all the IC outputs connected to a bus, while all the ICs are disabled measure and store the current value on the bus, then individually enable each device while measuring and storing the resulting current. When all the ICs have sequentially enabled a current value, a comparison is performed. Any IC output that causes a significant change in current is a good device. Any IC output that essentially remains the same is a defective device.

6.8. DIGITAL TESTING

Digital logic testing is the next level in the test hierarchy. The digital logic tests are performed by calling up a library subroutine that typically exercises the IC inputs and compares the IC outputs for all stuck-at-1, stuck-at-0, and pin-to-pin shorts faults. The test time typically ranges from 5 microseconds to 300 milliseconds per device.

To test individual ICs independent of the surrounding circuitry, it is necessary to have total control of the IC's input stimuli while the outputs are sensed for the correct response.

6.8.1. Backdriving

In 1968, Testline, of Titusville, Florida, now the Factron/Schlumberger, Titusville division patented a high-current, low-energy, short-duration, injection pulse technique to accomplish the backdriving task. The over-drive pulse isolated the DUT from the previous stage logic levels, and defined the input test vectors. In the upper portion of Fig. 6.4, if a logic 1 is injected at node 6 (one of the DUT inputs) and the output state of inverter A1 is also a logic 1, no significant current will flow. However, if a logic 0 is injected at node 6 while the inverter is in a logic 1 state, current will flow from the inverter to the test system digital driver. For TTL compatible devices, this current is defined as the output short circuit current, I_{os}, and is a specific parameter. If the inverter output state

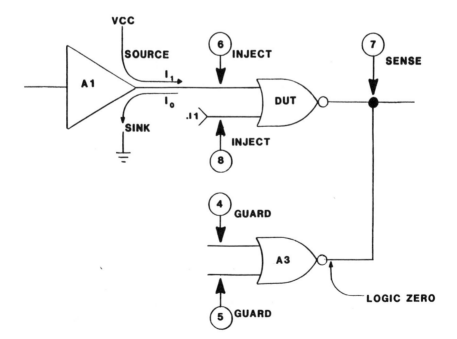

Fig. 6.4. Digital testing.

is quiescently at logic 0, and a logic 0 is injected at node 6, again no significant current will flow. Conversely, when the inverter output is at a logic 0 and a logic 1 is injected at node 6, current will flow from the test system digital driver through the output stage of the inverter as the device is forced out of saturation. This is because TTL logic is designed primarily for sinking currents. The current resulting from forcing a TTL output from a logic 0 to a logic 1 is generally not documented, but is usually estimated to be twice I_{os}.

An ECL device whose output is the emitter of a transistor, only experiences a significant current when an ECL output logic 1 is forced to a logic 0. This current is limited only by the output transistor gain unless a series resistor exists in the collector circuit.

CMOS devices are similar to TTL except that if the inject voltage pulse is above VDD or below VSS, SCR latchup occurs, which is essentially a VDD to VSS short.

For the purpose of clarity, backdriving is defined as the forcing of a device output to a state opposite to that defined by its internal circuitry. Backdriving is further divided into two categories: short pulses used for data bits, clocks, or resets; and long pulses used for sets, clears, and device enables. A question commonly asked is, what potential harmful effects both immediate and accumulative are caused by backdriving? Obviously the high-power, low-energy pulse causes a current in the output stage resulting in localized heating in the current-carrying transistor and in the metal conductors. If allowed to continue the heating will eventually spread to the entire chip and from there to the package body. Therefore, if a test system does not control the inject pulse energy, power, duration, and duty cycle within safe limits, the potential exists for immediate or catastrophic damage and/or accumulative degradation, which reduces component life.

Without going into the physics of an integrated circuit, the concerns are:

1. Thermal fatigue and metalization fusion occurring in bond wires, bead leads, or on the silicon chip itself
2. Secondary breakdown of junctions from hotspots which cause a temperature rise above the intrinsic temperature
3. Electromigration, i.e., the actual movement of the atoms of the chip's metalization caused by high current densities

4. Surface reconstruction, i.e., the destructive restructuring of aluminum metalization from thermal cycling

5. Electromagnetic degradation, i.e., the conductor connections stresses caused by high current densities

Over the past several years many stress studies have been conducted on the backdriving of TTL, ECL, and NMOS, and CMOS devices. Results show if a stress pattern is applied for more than 300 milliseconds with a 50-percent duty cycle at greater than 500 milliamps at 5 volts, or a single stress is applied at 750 milliamps for more than 2 milliseconds, it will cause harmful damage. Accumulated stress times to damage is approximately 3 seconds for 10 tests at 300 milliseconds each.

Most commercial digital in-circuit testers perform incremental stresses on ICs for one microsecond to 200 milliseconds, at a 50 percent or greater duty cycle with maximum current ranges from 250 to 500 milliamps at 5 volts or less.

To date there is no evidence of incidents either reported or experienced in which a commercially available in-circuit test system has caused a reduction in device life through backdriving. Several major manufacturers construct their test programs so that the test sequence begins at the end of a logic path and proceeds to the start point(s), thereby ensuring there have not been any immediate or catastrophic damage by backdriving the preceding stage.

6.8.2. Digital Guarding

By employing backdriving, IC input stimulus is under total control. However, the output response may be masked by common logic path activities. To isolate the DUT's output, digital guarding is employed. All IC inputs with outputs common to the DUT output are forced to provide a known-output-state while the DUT is being sensed.

In the lower portion of Fig. 6.4, as the DUT's logic is being tested, nodes 4 and 5 are being injected with a logic 0, maintaining a high impedance at the wired or output state. Similarly the digital guarding can be used to break feedback loops and disabled counters.

Most in-circuit testers offer a combination of digital testing capability; sequential static, parallel static, or high-speed parallel truth table testing with cyclic redundancy check (CRC). Each of these techniques is explained below.

6.8.3. Sequential Static Testing

Static digital testing is generally defined as a test rate less than 1 MHz. Most static digital in-circuit testers have a test pattern rate ranging from 10 to 100 kHz. Fig. 6.5 illustrates the digital pin electronics for sequential static testing. The five-bus matrix is connected to Family A programmable pulsed power supply, Family B programmable pulsed power supply, and the input to a dual-threshold sense amplifier. Each digital test point is connected by solid state CMOS switches to each one of the matrix buses. The test program closes the appropriate switch first to select the logic levels and second to determine

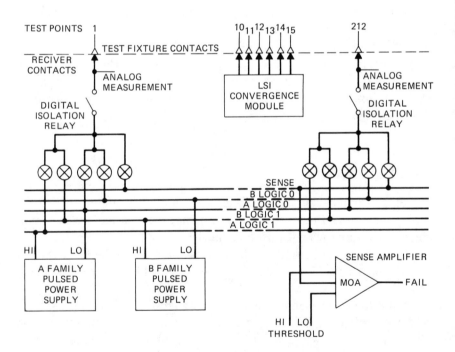

Fig. 6.5. Sequential pin electronics.

the sequence forcing the IC inputs and measuring an IC output. The digital isolation relay is employed to provide a hybrid test pin. If the test point is pure digital, the isolation relay is omitted. The pins' slew or delay time between switch closures is on the order of hundreds of microseconds. This greatly simplifies the decoupling problems caused by text fixtures, as no more than one driver makes a transition, and fast enough for SSI/MSI logic that the sequential drive inputs and the scan sensed outputs have no effect on the test quality. The dual threshold is programmed to an A or B Family high or low logic level. One pass or fail flag is determined per test step. This type of digital pin has the benefit of being very inexpensive.

6.8.4. LSI Convergence Module

When sequential, digital logic testing is used, in many cases support hardware is required to facilitate the testing of large scale integrated circuits. When support hardware is required, the specific circuit is typically installed in the test fixture head. In investigating the types of specific circuits that are commonly required, one will find single-step circuitry (pulse catchers), synchronizing device input signals, providing device inputs of short durations or other device signals, and providing an oscillator to a dynamic device-under-test if one is not available on the UUT.

Consequently, in an effort to eliminate or sharply reduce the requirements for specific circuits being installed in the test fixture head, for lack of a better name, the LSI convergence module was introduced. Four or five cards with a variety of parts and circuits – D flip-flops, latches, hand gates, inverters, clock circuitry, oscillator circuitry, synchronous one shots, pulse catchers, registers, and resistors – are installed on the opposite side of the system interface receiver. This support circuitry is interfaced to the DUT via relays that are included in the LSI convergence module. To access the LSI convergence module, the support circuitry wiring connections are made in the receiver and implementation is under software control.

6.8.5. Parallel Static Testing

Figure 6.6 illustrates parallel static pin electronics, commonly referred to as RAM back pins. The test program loads each digital pin's

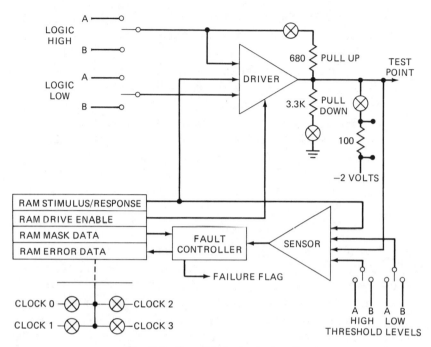

Fig. 6.6. Simplified pin electronics.

memory with the appropriate vector patterns for test execution. The stimulus/response RAM commonly referred to as the data RAM, contains both the logic 1 and 0 for the drive and sense logic. The drive enable determines whether the pin will be a driver. The sensor is always enabled in order to either monitor the driver's logic or the output of the DUT. The mask data RAM determines if the sense results that fail should be ignored, as when initializing a device, or should be stored in the error RAM. At the output of the driver, both pull-up and pull-down resistors may be incorporated by relay closures along with ECL termination. The test pattern rate ranges from 50 to 250 kHz depending on the RAM execution rate and total driver/sensor circuitry response time.

The advantage of parallel static digital pins is that no external circuitry is required when a device requires a large number of simultaneous drive or sensed outputs. Of the two static digital testing capabilities, sequential is the more popular because it meets minimum technical requirements at significantly less cost per pin.

6.8.6. High-Speed Parallel Testing

Figure 6.6 can be converted to high-speed parallel pins by adding a high-speed clock and control circuitry. Present-day high-speed parallel pins have a test pattern rate ranging from 1 to 5 MHz. High-speed parallel driver/sensors allow fast test program execution. This rate is adequate to test LSI devices without exceeding the safe backdrive time limitation. Test pattern length is a function of the RAM memory size unless a controller is included that allows looping, jumping, etc. At speeds over 2 MHz, exhaustive patterns may be applied to memory devices to further test for cell integrity and interactions without exceeding safe backdriving limits.

6.8.7. Signature Analysis

On static digital testers, memory and memory arrays are tested by the use of cyclic redundancy test or signature analysis (CRC). Signature analysis relies on a data compression technique for a serial bit stream. The data compression measurement and analysis function of signature analysis is an algorithm derived by selective feedback of the contents of a shift register, typically 16 bits, to which the data stream is applied. The signature interval and data are defined by start/stop gates and clock signals. This signature, four hexadecimal characters, represents the data stream at a node during a specific time interval. Reapplying the same stimulus under the same conditions to an IC input will result in a repeatable output signature unless the IC is defective. The probability that a valid signature would be duplicated accidentally by a bad serial stream is effectively the same as the signal bit error detection probability of 99.9985 percent. Conversely, the probability that a bad serial stream accidentally duplicates a valid signature is 0.0015 percent. In practice test vectors and resulting signatures are stored as library elements. In many cases, the preparation software allows the IC inputs to be stimulated while the output signature is learned.

6.8.8. High-Speed Data Compression

Another form of high speed pin electronics is illustrated in Fig. 6.7. An additional RAM has been added to control the latch at the output of the data, drive enable, and mask RAMS, plus the input to the error RAM. This RAM allows for test program data compression, as only changes in state sequence are required. The control RAM, driven by a sequence microprocessor, determines how many test steps a particular input/output is held in the same logic state. Under software control, the burst length of uncompromised data stream is extended beyond the physical cell or state capacity of the RAM memory device. Further, with the use of software routines, the test vector sequence may be altered in real time.

Remember that the only differences between an in-circuit tester's digital pin electronics and that of a functional tester lie in the last stages, drive and sense amplifiers. The in-circuit tester's digital pins source and sink are typically 500 milliamperes, while a functional tester's digital pins source and sink are typically 40 milliamperes.

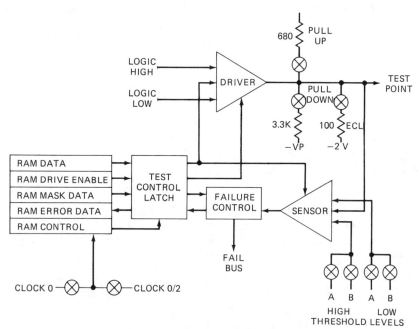

Fig. 6.7. Data-compressed high-speed digital pin.

The difference in current requirements is reflected in the difference in test rate. In general, the lower the source and sink current requirements are, the faster the pattern rate. The in-circuit tester's software philosophy is simplicity in both test program preparation and test program execution.

6.9. TEST PROGRAM PREPARATION

The PCB test kit, set, or package consists of a test program and a test fixture. The first step in generating a PCB test program is the

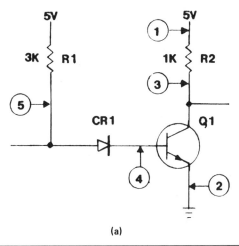

(a)

BUS	VOLTS	PIN
VCC	5.0	1
GND	0.0	2

RESISTOR	VALUE	TOLERANCE	PIN 1	PIN 2
R1	3K	10%	1	5
R2	1K	10%	1	3

DIODE	ANODE	CATHODE
D1	5	4

NPN	EMITTER	BASE	COLLECTOR
Q1	2	4	3

(b)

Fig. 6.8. (a) Nodalized circuit, and (b) data set.

preparation of the PCB circuit description. The PCB circuit description consists of a list of components along with their values, tolerances and nodal points. If a test fixture is not available or is being built in parallel with the test program development, a common method is to nodalize the schematic diagrams by assigning nodal numbers at each component interconnection. Fig. 6.8(a) illustrates a simple nodalized circuit. Each nodal point represents the respective test fixture test probe. Once the schematic diagram is nodalized and verified, it is a clerical function to format the circuit description into a data set that is acceptable to the automatic program generator (APG). Fig. 6.8(b) is an example of an analog data set of the nodalized circuit of Fig. 6.8(a). Fig. 6.9 shows an example of a digital data set.

Various utility programs and special routines help make test program development easier. One common utility, nodal learn, is employed when the PCB test fixture has been built from nodalized artwork. The test programmer enters the type, identification value, and tolerance of each analog or digital component. Then by software, via the VDT screen prompts, the programmer probes each analog component lead, or clip probes each digital IC. The nodal learn program automatically learns and stores, in the correct format, a data line for each component. The nodal learn program improves the data set reliability by eliminating errors in node number transfer, typing errors, nodalized schematics, inconsistencies between

BUS	VOLTS	PIN
GND	0	1
5V	5	7
12V	12	14
THRESHOLD HIGH		2.4V
THRESHOLD LOW		0.8V
DRIVE HIGH		4.5V
DRIVE LOW		0V
PULSE WIDTH		5us
CLOCK FREQUENCY		100KHZ
IC 7404	U3	41,18,11,15,36,34,L,63,20,19,25,18,48,H
LSI 8080	U35	95,L,23,61,66,65,87,85,64,60,U,68,28,62,32,43,70,11,17,H,!
		40,35,37,41,71,69,51,U,52,57,58,74,75,43,40,44,5,91,46
LSI 2102	V16	40,20,34,19,31,9,10,95,38,11,8,33,32,56,41,19
ROM 8016	V20	ADDR−54,40,36,70,35,52,56,82,83,40,58!
		DATA−48,51,30,56,53,57,50,35

Fig. 6.9. Digital data set.

schematics, UUT, and test fixture. Further this utility identifies fixture errors of test points not making contact, test points not wired, and missing test points.

Another method of obtaining a data set is from a computer-aided design (CAD) system which transfers the PCB parts and interconnection files into a data set generator. The data set generator assigns node numbers to each component junction. If a data set generator is not available, the nodal learn utility is employed as stated above. Before the data set is entered into the APG, it must first pass a full syntax check. The syntax check ensures proper routine, construction, and translation.

6.9.1. Automatic Program Generator

Referring to Fig. 6.10, the APG identifies a type of analog component and applies the appropriate analog test routine. The normal APG measurement analog subroutines are shorts/opens, resistors, capacitors, inductors, parallel networks, forced DC voltage measure current, forced DC current, measure voltage, forced AC voltage measure voltage, and differential voltage. Further, based on circuit construction, the test program algorithm will position the appropriate guard points to electrically isolate the component-under-test. The number of guard points is limited in the test program algorithm, typically 3–5. The number of nodes away from the component-under-test is also limited, typically 2–4. The analog test routine then incorporates the component-under-test value and tolerance, which results in a complete test program for the component-under-test. For any test that cannot be executed because of circuit configuration, a warning flag is set and the test is automatically ignored.

The APG identifies the digital IC and fetches the appropriate digital subroutine from the digital library. The APG then modifies the library subroutine to fit the circuit configuration and disables all IC devices and breaks feedback loops to digitally isolate the device-under-test. For example, a NAND gate has its inputs tied together. The test for pulsing the inputs to opposite states are not legal. The APG takes this into account and ignores the tests that attempt to drive the inputs to these states. Again, if a device is not testable due to circuit configuration, a warning flag is set and the test is automatically ignored.

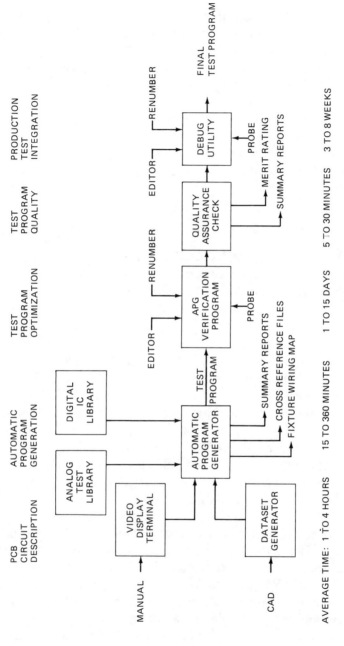

Fig. 6.10. Test program development.

6.9.2. Digital Library

The digital library consists of a series of test procedures, input test patterns, and resulting outputs that test for stuck-at pin faults for a particular IC device. The digital library contains SSI, MSI, some LSI, and possibly a few VLSI test routines for DTL, TTL, CMOS, IIL, NMOS, and ECL. Many in-circuit testers have problems with memories containing delays, dynamic LSI, and mixed logic. Many APGs incorporate an automatic learn mode for learning ROMs, PALs, and PROMs. The digital test execution is logic comparison employing truth tables for standard logic devices and truth tables or CRC for memory devices. Most digital libraries have the capability for the user to add custom elements.

The minimum test requirement of a digital library subroutine is to verify that all output pins can drive high and low to reasonable levels specified in the data sheet, independent of all other output pins, and that all input pins can sense high and low independent of all other input pins. For the output pins, this result may be achieved by setting conditions that all outputs other than the one in question are driving at one level, then forcing the output in question to the opposite condition using logical inputs on the input pins. To test the inputs, the same technique is used. All inputs are driven at one logic level and the outputs are tested for a known condition, then the input in question is driven at the opposite logic level and outputs are tested again. The operations are repeated until all input and output pins have been tested.

6.9.2.1. SSI and MSI Library Elements. When generating test source code for small and medium scale devices, it is possible to test all conditions and logic states for all elements (gates, flip-flops, etc.) of that device. Data book truth tables and math models provide, in most cases, an acceptable array of test conditions and logic states. The Boolean equation for a single element of 7400 would result in a truth table.

One of the first tasks a programmer must accomplish when coding an MSI device is initialization, forcing the device outputs to a known state. Clear and preset lines are undesirable for this task. Often these lines are inaccessible when found in-circuit.

Device test programming for MSI devices uses the same process. Multiple element devices should include a separate test sequence for each individual element. Each control line should have a separate test sequence that verifies its proper operation (e.g., clear, set, preset). If control lines are prioritized, a priority test sequence is necessary. The test that verifies the primary operation of the device should be written so minimum test requirements are met. For example, when creating the test source code for a serial-in/parallel-out shift register, one method for the primary test (shift right) would be to clock into the register an alternating pattern of zeros and ones (01010101) and then test the parallel outputs for this pattern. The inverse of this pattern (10101010) should be clocked into the register and the parallel outputs tested for this pattern. If a device has latched outputs, a new set of data should be loaded prior to testing the device outputs to ensure there is no corruption of the latched data. This example covers the minimum requirements for the input and output pins.

6.9.2.2. LSI Library Elements. There are two levels of testing for LSI devices. With lower-level tests, all control lines, interrupts, and service request lines should be tested along with exercising registers and executing major instructions. On a higher level, a complete test of the registers, instructions, and control lines would be initiated. Because of the complexity of some LSI devices, this higher-level testing is not always possible. Intended primarily to detect manufacturing faults, in-circuit testing looks for defects in a printed circuit board, including bent and broken pins, solder splashes, and devices that are "destroyed." Unlike component testers which attempt to measure total functionality, an in-circuit test will identify catastrophic rather than subtle failures. During the course of an LSI device test procedure, a number of internal registers and functions are exercised. The assumption is if the device is largely nonfunctional, it will fail.

The APG output is the PCB test program, a series of cross reference files, summary reports, and possibly a test fixture wiring map. An example of a test program statement is shown in Fig. 6.11. The test program is optimized by employing an APG verification utility. The engineering-oriented test program code is for a 1-kilohm resistor. Resistor 1's test has been assigned program number 100. If the test

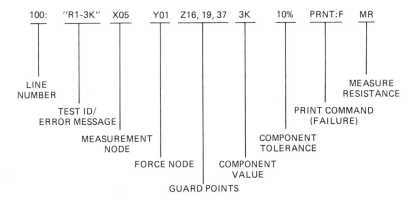

100:    "R1-3K"   X05   Y01   Z16, 19, 37   3K   10%   PRNT:F   MR

LINE
NUMBER   MEASURE
                                                                              RESISTANCE
        TEST ID/   PRINT COMMAND
        ERROR MESSAGE   (FAILURE)

            MEASUREMENT   COMPONENT
            NODE   TOLERANCE

                FORCE NODE   COMPONENT
                                  VALUE

                    GUARD POINTS

Fig. 6.11. Test program statement.

fails, "R1-3K" will be printed at the failure printer. Test points 5
and 1 have been assigned to the resistor's nodes. Guard points eliminating parallel current paths, thus isolating the UUT from adjoining
circuitry, are 16, 17, and 37. If measured beyond the 10 percent
tolerance assigned by the programmer, the resistor will be failed.

6.9.3. APG Verification

Many test systems have an APG verification routine where a known-
good-board is used to verify the test program. The APG verification
routine measures the device three times as directed in the test program
and stores each value, then reverses the leads and measures it three
more times. If there is a difference in value or stability, the most
accurate and stable test configuration is stored as a test program enhancement. Each guard is then removed in sequence and three
measurements are repeated looking for divergence from the nominal
value and stability of measurement. If removing any guard causes a
drift in nominal value or in stability, that guard is immediately replaced. Further, the APG verification software aids the programmer
in attempting to isolate the device for best and fastest measurements.

The automatic program generator fetches the library model of a
specific device, then modifies the library subroutine to conform with
how the UUT is configured in the circuit. The APG automatically
positions the digital guard points. Unfortunately, most automatic

program generators are not 100 percent efficient in dealing with complex logic containing multiple feedback paths. One would expect that some degree of debugging may be required.

6.9.4. Quality Assurance Check

Before test engineering releases a PCB test program to production, the test program is evaluated by a quality assurance check (QA) utility program. The quality assurance inspects the APG efficiency by comparing the generated test program with the original data set, then lists the test not included during automatic program generation along with a percentage of merit. The completeness of an APG varies with the in-circuit's APG algorithm and the complexity of the UUT. Generally the QA percentage ranges from 65 percent for very complex PCBs to 98 percent for simple PCBs before APG verification. After APG verification, above 90 percent is common for all PCB types.

6.9.5. Other APG Outputs

The cross reference APG files are aides in debugging the test program. The cross reference files typically consist of a component reference to test fixture probe numbers, test fixture probe numbers to component reference, interconnection wiring lists, parts lists with component test specifications. Summary reports include test not generated and components not tested. The test fixture test probe wiring map generated by the APG eliminates one source of human error along with reducing time and labor costs. This situation may not present a problem for low-population PCBs or in-circuit testers that do not multiplex test points. However, for high-population density PCBs and in-circuit testers with multiplexed test point fixturing, maps become a more critical item. The sequential operation of first generating the test program and then building the test fixture will increase the PCB test setup time. After APG the test engineer normally can choose a variety of utility programs and routines to help optimize the test program. The most important is the interactive editor.

6.9.6. Editor

The editor will position a cursor on the VDT screen where a change is to take place including character or line insertions and deletions.

Most editor commands are single key strokes which are executed immediately. The majority of editors have memory, which allows the operator to rearrange files and recover unintentional deletions. The actual time to edit and execute a test program is a function of the power of the editor and whether the in-circuit tester's software is a compiled/translator type or interpreter-based. In a compiled system the test program must be translated from objective code to source code for editing. Then after editing the test program must be compiled for test execution. In an interpreter-based system, the source code is edited and then executes the test.

The interpreter enables quick interactive modification of the PCB test program for increased debugging efficiency and accommodates frequent engineering changes. Interpreter-based software provides maximum system flexibility in a dynamic production environment. Further, an interpreter-based system offers more tangible representation of UUT test and increased diagnostic capabilities. The disadvantage of interpreter code is it runs more slowly than compiled code. However, an interpreter can be written so that the amount of time actually spent interpreting the code and branching off to the appropriate device becomes negligible, when compared to measurement setting time and test setup time.

6.9.6.1. Renumber. The renumber utility will automatically add and change test program line numbers. The programmer may insert a test statement between line numbers 100 and 101, or append two test programs that start with the same line number. The renumber utility accommodates these programming changes by automatically reformatting the test program line numbers to the programmer's specified instructions.

6.9.7. Debug

Another important utility employed to integrate the in-circuit tester into the production line is the debug program. The debug utility eliminates the need to search through a test program listing to determine the test parameters for a device. The debug display for analog components consists of error message, range of good values, test time, force, measure, guard points, test program line number, and code being executed and displays the actual message value. For digital devices the debug display consists of the error message, driver and sensor levels, pin translation table, digital guard points, test program line, and code being executed. The failing IC pin and failing test point number displays the actual measured and expected values.

6.10. IEEE-488 INSTRUMENTATION

The IEEE-488 interface bus is a standard feature on most in-circuit testers. The bus provides functional test capabilities that allow test program control of any programmable IEEE-488 compatible bus instrument and serves to minimize the use of dedicated hardware. Many in-circuit testers have a general-purpose interface bus software driver (GPIB) which enables stimuli and measurements using instruments under test program control. Software control is generally classified into two categories. The first is listen/talk, for instruments capable of receiving and transmitting data such as digital multimeters and frequency counters. The second is listen only, for instruments capable of receiving data but unable to transmit data, such as programmable power supplies and signal generators.

Most in-circuit testers have auxiliary relays under software control, consisting of 16 or 32 relays. The relay contact terminals are accessible in the system receiver and may be implemented by interconnection of the receiver points. Standard IEEE-488 instruments may be incorporated into the system through hardware and software control. The appropriate instrument input/output is routed to the system's receiver by coaxial cable, then routed to specific UUT points for source and measurement functions.

Vendors may offer a separate instrument switching matrix to facilitate incorporating instruments. Several vendors offer three or four standard instruments as well as space in the system to accommodate them. Others offer additional cabinets as an option.

7
IN-CIRCUIT TESTER

The challenge for the in-circuit tester is to achieve greater throughput with a higher yield. This challenge requires compatible hardware and software for fast test program generation and execution.

A simplified block diagram of an in-circuit tester's architecture is shown in Fig. 7.1. The computer subsystem interprets the test program and transmits commands to the switching and measurement subsystem. The switching and measurement subsystem then develops the proper test criterion routing the appropriate signals, at the proper time, to the correct UUT interface subsystem point. The UUT interface subsystem further routes the test configuration to the test fixture system which contains the UUT. The UUT's reaction to the test stimuli is passed through the test fixture system and UUT interface subsystem to the switching and measurement subsystem for measurement. The results of the measurement of each test are transferred to the computer subsystem for analysis, storage, and reporting.

7.1. COMPUTER SUBSYSTEM

The computer subsystem illustrated in Fig. 7.2 is the intelligence center of the in-circuit tester. The computer could be an OEM minicomputer, microcomputer, or an in-house-designed microprocessor base unit. Typically, the computer would be an OEM, 16-bit-word minicomputer with multiaccumulator architecture, real-time clock,

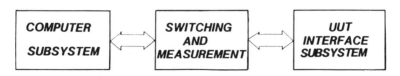

Fig. 7.1. Simplified hardware architecture.

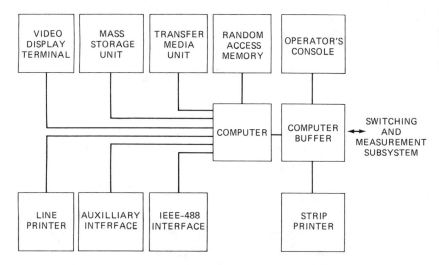

Fig. 7.2. Computer subsystem.

multiple inputs/outputs, and a memory management unit. The size of the resident random access memory is a function of the temporary storage requirements and how many activities are occurring simultaneously. Generally 256 kilobytes of memory are incorporated into most in-circuit testers minicomputers.

The transfer medium is the method employed to transfer data physically from one system to another, or from goods storage to a tester. The transfer medium may be paper tape, floppy diskette, removable cartridge disk, cassette tape, streaming tape, or a data communication link. The three most popular transfer media are dual-density, double-sided, floppy diskettes, five-megabyte cartridge disk, and RS-232 data communication link.

The mass storage medium is where all the test programs and data logging files are stored. The mass storage medium could be floppy diskette, hard disk, or Winchester disk. Generally a 20–75 megabyte Winchester disk is found on large in-circuit testers because of the Winchester disk reliability. On small in-circuit testers, the floppy diskette or small 5–10 megabyte Winchester disk is employed. The Winchester disk is 7–10 times faster than the conventional floppy disk and more reliable.

The video display terminal (VDT) is the primary input/output device to the computer and the in-circuit tester. The VDT is operator-oriented, with 24 lines of 80 characters on each line of the display, and an alphanumeric typewriter-style keyboard, producing 64 upper-case ASCII characters, and containing functional command keys and direct cursor positioning controls making data manipulation fast and easy.

The line printer, generally offered as an option, provides a hard copy listing of test programs and data logging. The line printer outputs at an approximate rate of 15 inches per second at six lines per inch vertical spacing, or 180 characters per second.

The auxiliary interface is a secondary input/output for background VDT, networking, telephone modem, or other data communication. The auxiliary interface generally has both RS232 current and voltage loops at 110–9100 baud.

The IEEE-488 interface is to compile programmable IEEE-488 instrumentation to the tester. Most in-circuit testers have general purpose interface software for listen/talk and listen only instruments.

7.1.1. Operator Console

The operator's console offers convenient tester operation by providing controls and communication at a single, easily accessible, location. The operator's console normally consists of alphanumeric visual display pass/fail status indicators. The typical system controls are main tester power on/off, start 1, start 2, pause, continue, single test, and end test.

The start 1 and start 2 actuate the test sequence and respective vacuum solenoid to actuate the right and left test chamber of the test fixture system. If only a single test chamber fixture is employed, one vacuum solenoid controls the test fixture vacuum. In this case, only start 1 control is required.

The pause control halts the test program execution. The continue control resumes the test program execution from the point where halted by the pause control.

The single test control allows the operator to step through the test program execution line-by-line. The end test control terminates the test program execution and resets the system to the test program start point.

The strip printer is used as a failure message output that the operator can attach and affix to a defective PCB. The strip printer is normally an impact or thermal printer. The most popular thermal printer is one that prints at a two-line per second rate. The strip printer is commonly contained in the operator's console. In addition, some in-circuit testers have a diagnostic center as part of the operator's console. The diagnostic center consists of an IC clip probe for auto-learning in the test program preparation, voltage probe to measure voltage at PCB nodes, a current probe for detecting bus shorts, a ground probe to verify the test fixture/test probe continuity, and various jacks for triggering scopes and logic analyzers during test

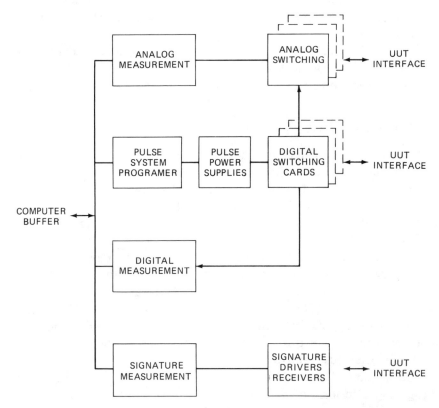

Fig. 7.3(a). Switching and measurement subsystem — analog/sequential digital with SA.

program debug or library element verification. The computer buffer is the interface to the switching and measurement subsystem.

7.2. SWITCHING AND MEASUREMENT SUBSYSTEM

Figure 7.3(a) illustrates a switching and measurement subsystem containing sequential digital testing, analog measurements, and signature analysis. As discussed in Chapter 6, the pulse power supplies, typically four (Family A voltage high, Family A voltage low, Family B voltage high, and Family B voltage low), are programmed to specific voltage levels determined in the test program. The digital switching cards, 16 or 32 digital channels per card, employing CMOS solid-state switches, route the appropriate logic signal to the correct test point on the system receiver. The UUT output response is routed through the system receiver to the digital switching card. The digital switching card routes, in test program controlled sequence, the logic response to the digital measurement unit for comparison. The digital measurements first measure pulse high/low sequence to a programmed dual threshold, a logic 1 and logic 0 voltage level. The resulting carry is strobed synchronously with the stimulus pulse. This type of measurement is generally used for combinational logic. The second is a "measured latch" which inspects a level at some time after the stimulus has been executed. This measurement is generally used in testing sequential logic.

In the analog channel, the test program directs the analog switching cards, 16 or 32 channels per card, to connect, employing a Reed relay, the appropriate analog measurement bus to the proper system receiver test point. To configure a hybrid test point each digital channel output pin is connected to the analog switching card via an isolation Reed relay located on each analog switching card. The signature analysis drivers and receivers are connected directly to the system's receiver. Typically eight drivers and eight receivers are included for memory array testing.

In-circuit testers employing sequential digital testing are less expensive per test point than parallel digital testing. The testers have a maximum number of digital and analog switching card slots. The maximum number of digital switching cards multiplied by the number of channels per card is the maximum number of digital test points.

The maximum number of analog switching cards multiplied by the number of channels per card is the maximum number of analog test points. The maximum number of hybrid test points is equal to whichever digital or analog test point number is the lowest. The remainder is pure analog or pure digital, whichever the case may be. An example is 16 digital channels per digital switching card and 32 analog channels per analog switching card. The maximum tester slot configuration is 60 digital and 60 analog switching cards. The tester configuration is 960 digital and 1920 analog test points or 960 hybrid and 960 analog test points. As an unofficial guideline, today's hybrid PCBs generally require 65 percent digital test points and 35 percent analog test points.

Figure 7.3(b) illustrates a switching and measurement subsystem containing parallel digital testing, analog measurement, and signature

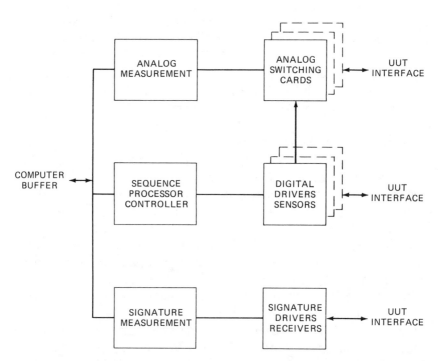

Fig. 7.3(b). Switching and measurement subsystem — analog/parallel digital with SA.

analysis.   As discussed in Chapter 6, each digital channel in parallel digital testing has individual RAM-backed driver/sensor circuitry. Sixteen or 32 channels are packaged on each digital switching card. The sequence processor controller incorporates a microprocessor, memory, and buffers to provide driver/sensor control such as nesting subroutines, conditional and unconditional pattern looping, and branching.   To other subroutines associated with the sequence processor controller is a driver/sensor programming control that transfers data, set-up logic family levels, and controls status and timing functions.   Generally efficient data transfer between the CPU, sequence processor, and digital pin memories is achieved through direct memory access, DMA, employing a 16-bit-wide data bus structure.   The analog measurement and signature analysis are the same as in the switching and measurement subsystem containing sequential digital testing, analog measurement, and signature analysis.

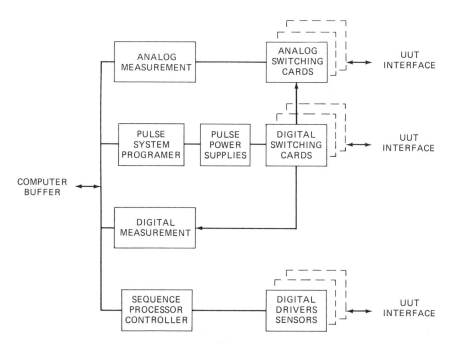

Fig. 7.3(c).   Switching and measurement subsystem — analog/sequential digital/ parallel digital.

Figure 7.3(c) illustrates another switching and measurement configuration of sequential digital testing, parallel digital testing, and analog measurements. The parallel digital testing may be high speed, greater than 1 MHz, where signature analysis would be less effective in testing memories and LSI devices.

7.2.1. Digital Pin Electronics

Sequential static digital pins operate at a pattern rate of about 100 kHz, with an uncompromised pattern depth of about 65 vector sets. Uncompromised pattern depth is the number of test vectors that are being transmitted to the UUT without any break or pause in the test vector rate. Uncompromised pattern depth has more meaning when applied to RAM-backed pins. Testing electronics using parallel static RAM-backed pins operate at pattern rates of about 225 kHz with an uncompromised pattern depth of 250–1000 vector sets. That is, after 250 or 1000 test vectors, a time delay is required to reload the RAM to be able to execute another burst of test vectors. Parallel high-speed (in excess of 1 MHz) RAM-backed pins with a microprocessor sequence controller have a pattern rate of 1–5 MHz with an uncompromised pattern depth in excess of 1000 vector sets. The digital drivers consist of two families, A and B, both highs programmable from –2 volts to +7 volts and both lows programmable from +2 to –7 volts, with a 50 millivolt resolution and 100 millivolt accuracy. This allows testing of TTL, CMOS, and ECL devices. In this case, the CMOS is limited to an even volt swing rather than a 12 volt swing. The driver output resistance is about 1 ohm with a capacity of sinking or sourcing at least 350 milliamps. The current injection has the ability to source at least 350 milliamps and sink at least 300 milliamps to ensure the capability of backdriving line drivers, level converters, and bus drivers. The sensors are also two-family (A and B) independently programmable high and low thresholds ranging from –6 to +6 volts, with a 20-millivolt resolution, a 50-millivolt accuracy, and an input impedance on the order of 5 megohms.

High-speed testing generally requires multiple clock phases. Timing and clock units supply either two or four clock phases and timing controls to simulate the requirements of the unit-under-test.

For a solely signature-analysis-type in-circuit tester, the driver and receiver are separate. There are commonly 32 pins per card. Typically twice as many receivers as drivers are required to test a PCB. A signature analysis is not only used for digital device testing. It is also used to functionally test blocks of digital logic where receivers are inserted along with logic paths for higher diagnostic visibility. For today's market, a maximum of 128 drivers and 256 receivers appear to be adequate to test most PCBs.

7.2.2. Analog Measurement

The analog stimulus and measuring unit is either a bridge-type or operational amplifier construction, as previously discussed in Chapter 6.

Typical analog specifications include shorts and opens, program crossover of 10 ohms to 100 ohms at an accuracy of 2 percent. Resistance ranges from 10 ohms to 10 megohms at ±1 percent. Capacitance ranges from 100 picofarads to 1000 microfarads at ±2 percent. Inductors range from 10 microhenries to 10 millihenries at ±5 percent. DC voltage range is 10 millivolts to 100 volts ±1 percent. DC current range is 100 nanoamps to 100 milliamps ±1 percent. AC voltage range is 10 millivolts to 71 volts RMS ±4 percent. Differential voltage from 1 millivolt to 10 volts at ±5 percent on the low end to ±1 percent on the high end. Source voltage ranges from 10 millivolts to 100 volts at ±1 percent. Source current ranges from 10 microamps to 100 milliamps ±1 percent. These analog specifications cover about 90 percent of today's PCB production testing requirements. The other ten percent includes a larger measurement range and better accuracy.

7.2.3. Switching Matrix

The switching matrix is essentially a big relay switch that connects various UUT components to the analog measurement unit or the digital pin electronics. Typically analog switching cards utilize highly reliable Reed relays packaged into 16 or 32 pins per card. The system's architecture determines the maximum pin count. A hybrid test point has a digital isolation Reed relay to remove the digital test electronic from the circuit when performing analog measurements.

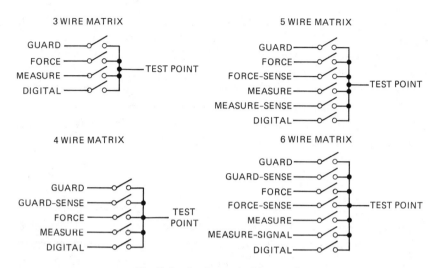

Fig. 7.4. Analog switching cards.

The analog switching matrix, as illustrated in Fig. 7.4, is defined by the number of wires per point that are connected to the measurement unit. Three wires allow force, measure, and guard functions. Four wires are force, measure, guard, and guard sense. Five wires are force and sense, measure and sense, and guard. Six wires are force and sense, measure and sense, guard and sense, as previously discussed in Chapter 6.

Each type of analog switching matrix has advantages and disadvantages. The three-wire matrix is simple and fast with no loop errors and no extra pins. Its fixture wiring is programmed independently; however, it tends to be guard-ratio sensitive and it delivers compromising accuracy at the low and high value. A four-wire matrix offers high accuracy at the high values, good accuracy with one guard, and a good guard ratio. However, it has low accuracy at low values when multiple guards are employed. In many instances, extra fixture pins are required and the program must generate a fixture map. The five-wire switching matrix is not presently employed in today's commercial in-circuit tester as it does not increase the guard ratio. The six-wire matrix is accurate throughout the range of components and offers an excellent guard ratio, but it has its disadvantages. Among them are the complex measurement systems and a requirement for extra pins. The test program generation must proceed before the fixture mapping can be derived.

7.2.4. Multiplexing

Multiplexing is a valid method of expanding a test system's point count. The true system test points are multiplexed a number of times to perform a large apparent number of UUT test points. One requirement to avoid a software dilemma is to have all the multiplex points the same; all hybrid, all analog, or all digital. Only the true points are connected to the system for UUT testing. Therefore, the UUT must be partitioned for testing. To maintain testing continuity, partitioned overlays are required, resulting in a number of redundant virtual test points. The number of duplicated test points may vary from 5 to 20 percent depending on the UUT's topography and the number of times the true test points are multiplexed. Fairly standard levels of multiplexing is times 2, 4, 6, or 8. Other serious restrictions come into play with excessive multiplexing. Multiplexing points is more of a software problem than a hardware problem. The accuracy of the PCB's topography is extremely important since the test program must be generated before the test fixture map can be determined. Early-life relay contact resistance failures, poor contact repeatability, and increased crosstalk leakage and transmission line reflections are all considerations.

The advantage of multiplexing points includes large point count capability, the achievement of a fully hybrid system, low cost per point, and low initial cost of ownership. The disadvantages of multiplex points are long startup time, more time to incorporate engineering changes, low mean time between failures, high mean time to repair, higher fixture cost, and lower product throughput.

7.3. UUT INTERFACE SUBSYSTEM

The last subsystem shown in Fig. 7.5 comprises the interface receiver, fixed and programmable power supplies, IEEE instruments, auxiliary relays, and for sequential digital testers, a LSI convergence module.

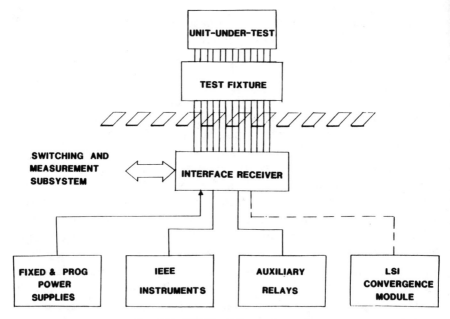

Fig. 7.5. UUT interface subsystem.

7.3.1. Interface Receiver

The interface receiver provides a reliable, easy-to-operate interconnection between the tester and the test fixture system. Most of the interconnections are from the switching and measurement subsystem. All the testers' analog and digital testing is accomplished through the interface receiver. The interface receiver may be a plane of spring-loaded, high-density contacts to a series of high-pin-density connectors. Generally all the test points available in the system are prewired to the interface receiver independently of the test capability ordered initially with the tester. That is, test point expansion is accomplished by adding the appropriate analog switching cards and/or digital driver/sensor cards.

7.3.2. Instrument Switching Matrix

The instrument switching matrix usually consists of 16 or 32 relays which are under software control. The relay terminals are accessible in the system's receiver and may be incorporated by interconnecting of receiver points. Standard benchtop instruments may be incorporated into the system through the hardware and software controls. The appropriate input and output are routed to the system's receiver by coaxial cable. The input and output are connected to the auxiliary relays that route the appropriate instrument to the UUT under software control. Some vendors may offer a facility of incorporating instruments directly into the system offering three or four standard instruments as well as space in the system to accommodate them. Typically any IEEE-488 instrument can be incorporated into the system. Added cabinetry usually is available as an option.

7.3.3. Power Supplies

Most vendors offer a group of standard UUT power supplies, plus a series of additional optional UUT power supplies to fit individual needs. The common standard power supplies are 5 volts at 15 amps, 12 volts at 5 amps, dual 12–15 volts at 3 amps, and 0–60 volts at 15 amps. Some test systems have a very desirable feature: when not testing a product, they self-monitor both the UUT and internal power supplies. If a power supply goes out of tolerance, a warning light on the operator console illuminates, and the test system halts until the power supply is recalibrated or a test engineer overrides the alarm.

The majority of in-circuit testers are equipped with vacuum test fixture capability. The vacuum pump, to isolate the noise source, is either remote from the test system area, or installed in a noise suppression cabinet. The vacuum is piped to the tester dual vacuum solenoid valves. The solenoid valves are actuated by depressing the start test buttons on the operating console. In many cases a pump off/on button mounted on the operator console remotely controls the vacuum pump's power.

7.4. GENERIC IN-CIRCUIT TESTER

Let us assemble the in-circuit subsystem into one simplified block diagram as shown in Fig. 7.6. The computer controls the UUT power supplies, analog stimulus and measurement, switching matrix, digital pin electronics, and IEEE, interface/instrument switching matrix.

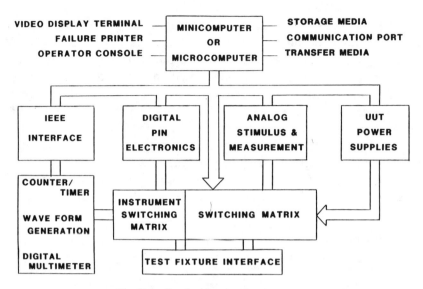

Fig. 7.6. Production in-circuit tester.

The digital pin electronics, analog measurement unit, and IEEE-488 instruments are all routed through the switching matrix to the system receiver. Most in-circuit tester manufacturers have tight quality control standards and well engineered systems. The mean time between failures in the industry ranges from 200 to 1000 hours. The mean time to repair ranges from 1 to 4 hours. The three factors that constantly appear to weigh most heavily are the test fixture probes, analog switching card Reed relays, and the mass storage drive reliability. There is a large variation between a Winchester disk, a removable cartridge, and a floppy disk drive. The mean time to repair is usually a function of the system's architecture, diagnostic aids, and availability of spare parts. When inquiring about a vendor's mean time between failures and mean time to repair, be sure you determine if the number is actual or estimated as well as what exceptions have been made. Obviously, actual data are preferred over an estimate.

7.5. OPERATING SOFTWARE

The system's operating software is essentially divided into three categories: (1) the operating executive, which is a function of the central processor's architecture and instructions; (2) the test executive, which converts the CPU into the system controller; and (3) the utilities, which operate under the test executive and operating executive.

Under the operating executive, we are interested in the type of language that is required to talk to the CPU, be it Fortran, Basic, or some other computer language. How is the storage medium constructed? Are the data overlayed, blocked, or linked? Does the computer have the capability of running multi-user — that is, testing a PCB and at the same time programming a different PCB? What is the capability of managing CAD/CAM data as an aid in developing the data base for automatic program generation? What variables are monitored internally and what flags are raised? What is the networking capability for up- and downloading with a host computer? What computer peripherals may be added to the system?

7.6. TEST EXECUTIVE

The prime question today is whether the test executive and/or library elements software is portable. This refers to whether the computer is essentially invisible, or common architecture. Supposedly this can be accomplished by writing the executive in Pascal. To date, no vendor offers a portable operating test executive.

Another question: What are the test program controls and debugging facilities? Test program controls include commands such as single-step, halt on fail, continue, pause, etc. When a failure occurs, what does the failure display field consist of? Does the display field have to be decoded — such as R1004(+) which means that R4 failed high? Is the same failure message that is displayed also printed out?

What software drivers for IEEE-488 plus applications are included in the test executive? How are instrument software drivers developed? Does the test executive have a general purpose IEEE software driver allowing the user to build his own instrument controls by adding the appropriate addresses and listen/talk codes for a specific instrument? Finally in digital logic testing, does the test executive include signature analysis, cyclic redundancy check, transition counting, truth tables, or function digital tables as far as stimulating and measuring device-under-test? Or is a combination of stimulus/measurement offered?

7.6.1. Utilities

The most important utility is the editor. It is an option of character, page file, or microinstruction editing. Is there the capability of logging data and then analyzing the data on the CPU? What about automatic test program loading when a fixture is placed on the tester? The appropriate program is loaded from mass storage ready to be executed. The self-check and fault diagnostics of the tester are essential. Is there the capability of putting a calibration adapter(s) on the tester and then executing a calibration routine that verifies the specifications of the measurement and driver/sensor sections of the tester? Fault diagnostics involve the capability of localizing a malfunction in the tester to an appropriate board or component.

7.7. TEST PROGRAMMING

The in-circuit software includes a hybrid automatic test programming generator. The programmer enters the PCB parts, both analog and digital with the respective node number, and the automatic test program algorithm determines the test configuration, test sequence, and guard points. The test programming generator outputs the test program quality rating, components not being tested, and an interconnection listing. The recommended categories for evaluating an in-circuit test programming generator are:

1. General implementation language, run time, quality rating, warning statements, and editing capability
2. Circuit modeling, automatic handling of fixture verification, IC orientation, bus faults and wired ORs, RC time constants, plus automatic disabling digital feedback loops
3. Analog measurement — number of measurements per capacitor, diode transistor, guarding capability; isolation of parallel networks and high- and low-value measurement capability; transition from measurement to test criteria for calculated values of capacitors, tolerances, and offsets
4. Digital measurement — fault analyzing techniques, fault states, fault resolution, and circuit complexity

5. IC modeling – number of ICs and what unique ICs, modeling techniques, ease of modifying, LSI capability, and self-learn capability

6. Automatic program generator control capabilities, system maturity, and interactive capabilities

The test program is verified and debugged on the test system by testing several good boards with the data log function in operation.

7.8. TEST PROGRAMMING STATION

A test programming station is a standalone computer system with mass storage, high-speed printer, transport medium, and multiple programming console. In many cases the test programming station will have the test electronics to debug the test programs and verify new library elements. The need for a programming station is often short lived, lasting as long as it takes to get the test operation under way and running smoothly. Typically a large number of PCBs must be programmed in a short period of time. After this only engineering changes are required. Therefore, the cost justification is sometimes difficult. Most in-circuit testers have test programming capability resident in the test system, with multi-user capability. Multi-user capability allows a programmer to generate a test program in the background while the test system is testing the product in the foreground. About 71 percent of manufacturers generate their test programs on the tester; 25 percent use a programming station, and the remaining 4 percent generate the test program manually. However, the only means of debugging a test program is on a test system with a series of known-good-boards.

7.9. BARE-BOARD SHORTS AND CONTINUITY TESTER

In some cases users wish to employ their in-circuit tester as a bare-board shorts and continuity tester. Of course, this is acceptable because a bare-board shorts and continuity tester is essentially a stripped down in-circuit tester with high point-count capability. Bare- or unloaded-board shorts and continuity testers are divided into two categories: high voltage (approximately 1500 volts) and low voltage (between 10 to 28 volts). A loaded-board shorts tester is a low-voltage, bare-board, tester except that it has a voltage excitation of 50–100 millivolts. The high-voltage tester employs Reed relay matrix modules, while the low voltage tester employs solid state switching modules. Therefore the test rate of a low-voltage tester at approximately 900 microseconds per point is much faster than a high voltage tester at approximately two milliseconds per point.

The bare-board shorts and continuity tester controller typically is a mini- or microcomputer with 128 kilobytes of memory and dual floppy or Winchester disk drive. The programmer's console is a video display terminal. These systems have self-learn capability. That is, a good board is placed on the test fixture and the system's APG develops the test program at a learn rate of about 90–180 milliseconds per mode. Some systems also have UUT test fixture verification routines to ensure the PCB/test fixture contact. Most systems have data log capability with an RS232 interface. The operator console has system status indicators and power switch start/stop control switches. The ticket printer is generally a thermal strip printer and the diagnostic center is generally a probe for pointing continuity verification. The high voltage relay modules come in 16, 32, 64, and 128 points per module. Solid state switching comes in 32, 64, and 128 points per module. Most systems offer an expandable-rack cabinet capability. Therefore, maximum point-count capability is a good question to ask.

7.9.1. Bare-Board Test Fixtures

The bare-board fixture usually requires 3–10 times the number of probes for an in-circuit tester, because a probe is required at each feedthrough hole to perform the continuity test. The loaded-board test fixture is one probe per node.

The test head adapters are mechanical, pneumatic, or vacuum. Mechanical and pneumatic adapters either force the probe field onto the board, or more likely, force the board onto the probe field. The vacuum-actuated fixture employs a plexiglass lid to form a vacuum seal over the board. The pneumatic standalone fixture appears to be favored for very large probe count (test heads above 20,000 points), and the mechanical standalone fixture on small-probe-count test heads (300 points or less). The vacuum fixture covers a range where it shows performance and cost related benefits. Maximum flexibility is attained by applying a universal probe matrix of 100-mill or 50-mill grid with a number of masks, actuated by either mechanical, pneumatic, or vacuum source. Some test fixture manufacturers offer modular probe matrixes that can be installed in the field.

7.10. TEST FIXTURE SYSTEMS

Generally speaking, test fixture systems are classified by the method of actuation − manual, mechanical, vacuum or pneumatic. Each type of fixture system is then subdivided into fixed or interchangeable test heads and fixed or removable probes. The vacuum fixture with interchangeable test heads and removable (snap out) probes is used almost exclusively by the in-circuit market. The board-under-test is seated on the probe field by a vacuum, drawn either by a neoprene rubber diaphragm, a movable top plate, or a fixed top plate with flexible gasketing. Each type has its own advantages and disadvantages for different applications. The vacuum is drawn from the test head, either by a hose or a manifold.

The probe density is a function of the probe pressure, normally measured at 75 percent of its travel with the vacuum pump employed. Typically, a vacuum pump rated at 30 CFM at 18 inches of mercury will accommodate 15 probes per square inch, using eight-ounce probes, or 30 probes per square inch using four-ounce probes. A vacuum pump rated at 41 CFM at 29 inches of mercury can accommodate 20 probes per square inch using eight-ounce probes and 40 probes per square inch using four-ounce probes. The typical maximum probe travel is one-quarter of an inch. When a board has some obstruction on the track side, a fixed-probe removable test head vacuum fixture may be employed with probes that have one-half inch of travel. Probes with one-eighth inch of travel are typically used for bare-board testing. The registration of the PC board-under-test is accomplished by the tooling pins.

7.10.1. Estimating Probe Count

A method of estimating the probes required for a particular loaded board (node count) is:

Discrete	*Quantity*	*Total*
2 Leg	2	
3 and 5 Leg	X 2	
5 and 6 Leg	X 3	
7 and 8 Leg	X 4	
9 and 10 Leg	X 5	
11 and 12 Leg	X 6	
ICs		
14 to 16 Leg	X 6	
LSI		
Number of Active Pins	2	
Total number of Nodes		
If Bused	−5% =	
If Not Bused	+10% =	
	Total	

Statistically, this method produces a nodal count accuracy within ±15 percent. The only sure method is to count the individual nodes on the mylar.

Probe sets for bare boards cost from $0.50 to $0.65 as compared to a loaded-board probe cost of $1.25 to $1.50 for 0.100-inch centers, $2 to $2.25 for 0.050-inch centers, and $20 to $30 for 0.010-inch centers. Drilling the plate costs from $0.20 to $0.30 per hole.

7.10.2. Contract Fixtures

When contracting to have a test fixture built, the customer should supply the contractor with: (1) one positive set, full scale one-to-one mylars of the PCB artwork for probe placement (NC Drill Tape); (2) one bare board for probe field verification; (3) one loaded board for selecting the proper probe tips and gasketing; (4) one assembly drawing of the PCB for layout; (5) one mechanical drawing of the PCB giving pertinent dimensions, locations, and sizes of tooling holes and their respective tolerances for layout; and (6) one board schematic for nodalization. The majority of fixture vendors are mechanically oriented and are well versed in talking about contact pressure, number of probes per square inch and probe repeatability. However, the majority of vendors have an inadequate knowledge of the electrical concepts involved. This is not to imply that knowledge of mechanical fixtures is not essential. Typically, a brand new probe has a contact resistance of 10–15 milliohms, as a function of its composition. The mechanical probe reliability is typically in millions of operations. This is a nebulous term because most probe fatalities are caused by degradation of the probe head. The probes are grouped by the geometry of the probe head, such as pyramid, point, serrated, crown, etc. The ideal choice is a self-cleaning probe with multiple sharp edges to penetrate any flux or contamination on the bottom of the board.

7.10.3. Dual-Chamber Fixtures

The cost effectiveness of a dual vacuum-chambered fixture is the elimination of operator handling time. Typically, an operator takes four seconds to load a PCB on a fixture and actuate the vacuum plus six seconds to ticket and unload it, for a total handling time of approximately 10 seconds. A dual vacuum-chambered fixture consists of two independent vacuum chambers in a single fixture. As the test system is testing the PCB on chamber 1, the operator is ticketing, unloading and reloading the PCB on chamber 2. Similarly, while the test system is testing the PC board on chamber 2, the operator is ticketing and unloading and reloading the PCB on chamber 1. This ping-pong activity is in parallel with the PC board test time, making the effective handling time zero. The requirement is that the test system have two vacuum-controlled solenoids, so that the vacuum source need not be switched between the two separate vacuum chambers.

7.10.4. Multiple-Well Fixture

A multiple-well fixture has multiple probe fields and gasket silhouettes for identical boards or different boards. The advantages are: (1) the capability of sequentially testing a number of identical boards on the same fixture; (2) the capability of sequentially testing a number of PC boards that are interactive to form a product or subassembly; (3) one individual test for a number of different PC boards on a single fixture.   The unused wells would be covered with a plastic cover.

7.10.5. Fixture Kit

Most fixture manufacturers offer kits to build PCB fixtures.   The user drills the fixture, and inserts the probe receptacles, wire, gasket, etc.   These kits reduce the typical contracted finished fixture turn-around time of 4–6 weeks to 1–2 weeks.   In many cases the customer's own NC drill tape or the bare PC board will serve as a template.   Many manufacturers send the fixture plates to the manufacturer of their PC bare boards for drilling.

7.10.6. Universal Fixtures

The advantage of universal fixtures is reduction of fixturing costs and test-head storage.   The disadvantage is that if the main portion of the universal fixture malfunctions, the test system is down and the high initial cost is equivalent to 4–10 individual custom test fixtures.
   There are four basic approaches to universal fixturing:

1. Probe matrix on 0.100-inch or 0.050-inch grid and a number of masks to inhibit unused probes from making contact with the UUT. The advantage is that only one test-head is required and the cost of masking is very low. Disadvantages are that it requires an unusually high point count in the test system and decreases probe life because of the fatigue from being inhibited.
2. A removable platen on a probe matrix on 0.100-inch or 0.050-inch grid with a number of UUT gasket plates and probe field configurator heads. The advantages are a lower system point count requirement, a reduction in the test head cost as compared to Item 1, and an increase in probe reliability. The disadvantage is that every PCB requires a test gasket plate and probe field configurator test-head.
3. A removable platen on a probe matrix, 0.100-inch or 0.050-inch grid with a number of masks to inhibit unused probes and a number of probe field configurator test heads per PCB

family. The advantages are a lower test system pin count requirement and, compared to Item 1, an increase in probe reliability. The disadvantage is the requirement of a probe field configurator test-head for every PCB family and a mask for every PCB.

4. Card personalizer. This is not a probe matrix, but a test-head containing a probe field which contains a large number of common probe points from a number of PCBs, and a number of masks to inhibit unused probes. While a card personalizer is a very cost-effective method of universal fixturing, the disadvantage is that each PCB requires a corresponding gasket/mask.

7.11. PCB LAYOUT GUIDELINES FOR TESTABILITY

1. Standardize on trace/pad spacing. The PCB should be laid out on a 0.100-inch grid. In any event, trace/pad spacing should be 0.100-inch or greater. Microchip type PCBs have the PCB laid out on 0.050-inch probes and 0.100-inch probes can be intermixed but they cannot be interchanged. In addition 0.010-inch probes for limited functions are also available.

2. For reliable PCB probe field alignment two or more tooling holes should be provided as part of the board drilling pattern. The tooling hole should be positioned asymmetrically near a corner of the PCB and uniquely for each PCB type to ensure that the correct test head is employed in testing the PCB. The tooling holes should have a minimum diameter of 0.125 inch with a ±0.002–0.003-inch tolerance. Their position on the PCB should have a 0.002-inch tolerance. When 0.050-inch probes are employed, the tooling holes should have a diameter tolerance of 0.001 inch and a position tolerance of +/–0.002 inch.

3. To ensure an adequate vacuum seal, allow at least 0.187 inch unpopulated area around the PCBs periphery unless edge damming is used.

4. Solder-fill all rivet holes and open holes to avoid using gasketed plugs or vacuum covers.

5. Provide a 0.030–0.080-inch round or square test probe pad on each trace to permit uniform electrical contact. For 0.050-inch probes provide 0.030 inch or greater test pad.

6. To avoid special fixturing costs, ensure that every node or trace can be probed from the solder side of the PCB. This includes unused nodes.

7. Socket all larger pin ICs (20 pins or greater) or low-reliability chips for easy insertion and replacement.

8. Protruding leads should be trimmed to 0.09 inch or bent at a 90 degree angle parallel to the solder side of the PCB.

9. Avoid mounting any component or wires on the solder side of the PCB. Extending more than 0.187 inch may cause problems unless the fixture has been specially designed with a gasket pocket to accommodate the part. Generally the added fixture cost and gasket pocket reliability are not worth the effort.

10. Ensure the solder side of the PCB is free from solder flux and any contaminants that may impede solid electrical probe contact.

11. Arrange the components on the PCB to avoid high-density probe field areas and pockets that may collect dirt or trap moisture.

12. Clearly label all components, particularly ICs, and orient them in the same direction for each of assembling and detection of mix-orientation.

13. Mount all adjustable components for easy operation access. Adjustment controls should face upward or unobstructed from the closest PCB edge.

14. Use a solder mask whenever possible and keep the solder process constant.

15. Use a consistent type of PCB base and track material from board to board.

16. Whenever possible, standardize on connect pin assignments, power, input, output, etc., and provide a ground track around the entire PCB periphery.

8
IN-CIRCUIT TESTERS FOR
SERVICE AND REPAIR

The five basic service repair strategies should be reviewed prior to any discussion of in-circuit tester applications in service.

8.1. SERVICE REPAIR STRATEGIES

8.1.1. Throwaway Repair

When module repair cost far exceeds new module cost, the defective module is thrown away. For example, throwaway repair strategy is found in consumer industries for stoves, washing machines, dryers, TV sets, radios, and automobiles. The philosophy for the throwaway module repair is to have a distributor wherever a large product concentration exists. This distributor would supply modules both to local, company-owned service and third party service concerns, your friendly TV repairman, or local filling station.

Throwaway repair strategy advantages are: immediate customer satisfaction, fast repair time, guaranteed repair quality, medium-skilled technical manpower, and low training cost. The disadvantage: a large spares inventory. The throwaway repair strategy has no application to PCB automatic test equipment.

8.1.2. Board-Swapping Repair

Board-swapping repair entails exchanging a good board for a defective board in the customer's product. This is the most widely used repair methodology in the computer, computer peripherals, telecommunications, industrial, and aerospace industries. Implementation of the board-swap repair philosophy requires strategically located repair depots or regional offices with a captured inventory. These offices would both support and control a network of local or area service centers and field service engineers.

Advantages include customer satisfaction through fast repair time, guaranteed repair quality, medium-skilled technical manpower, and low training cost. Disadvantages are: large spares inventory, high shipping and handling cost, large board float, and excessive dependence on management. The majority of PCB testers use board-swap repair strategy.

8.1.3. On-Site Repair

This repair philosophy entails having a field service technician repair the defective printed circuit board on-site. This repair strategy is rare except where one printed circuit board is the total product, or the printed circuit board must be matched or tuned to other system circuitry. The on-site repair philosophy requires having a network of field service engineers dispatched by strategically located offices.

The disadvantages of this strategy are customer dissatisfaction because of long repair time, high-risk repair quality, and the requirement for a highly trained, sophisticated technician.

8.1.4. Walk-In Repair

Having a customer bring the defective product to a local service office for repair is defined as walk-in repair. This strategy is employed in sophisticated consumer items such as telecommunications, computers, and peripherals.

The advantages include a quality product and satisfied customers through feasible work load, low-skill technical manpower requirements, low operating cost, and reasonable turnaround time. The disadvantages are high initial setup cost, large spares inventory, high personnel turnover, and high training cost. Walk-in repair imposes a heavy requirement on PCB automatic test equipment.

8.1.5. Remote Diagnostics

Via a computer terminal and modem, the local center service technician diagnoses the customer's failure to board level. After identifying the customer's defective PCB, the service center either dispatches good replacement PCBs for the customer to swap, or dispatches a service technician with the proper PCBs for system repair. This strategy is widely used in the aerospace and computer industries. The remote diagnostic philosophy essentially is the same as the board-swap philosophy except each depot or regional office would contain a service communication area referred to as a "war room." This area would house the elements necessary to dispatch spare parts and local field service engineers.

8.2. SERVICE PROBLEMS

Numerous AFSM service managers surveys have revealed, as illustrated in Fig. 8.1, similar areas of concern: availability of skilled manpower, high cost of service calls, board float inventory size, budget allotment for capital equipment, and availability of proper ATE.

In excess of 80 percent of survey respondents identified availability of skilled manpower as the major concern. By incorporating automatic test equipment, the amount of skilled manpower required is sharply reduced. Field deployment of ATE reduces the overall service costs. This may be passed on as a reduction in service call costs. By far, the deployment of ATE in the field has the most significant

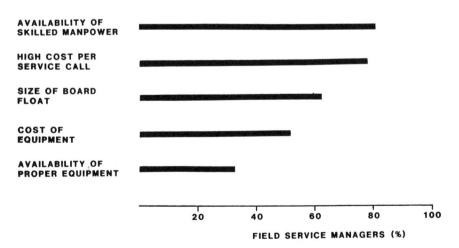

Fig. 8.1. Service problems.

effect on reducing the size and cost of the spares stock and board float. This will be discussed in detail.

The cost of test equipment is a two-edge sword. The service concern is generally defined as a labor intensive organization rather than a capital intensive organization. The requirement to repair high technology products is causing a shift in the service organization's philosophy. In addition, new technology is sharply reducing the cost of service test equipment. The availability of the proper test equipment has improved during the past two or three years because more companies believe field service is a viable marketplace.

One concern resolved by deploying ATE in the field is the size and cost of the spare boards inventory through a reduction in inventory carrying costs as well as shipping and handling costs. The problem remains despite a major effort being made to manage and control the spare parts inventory in the factories, the rework centers, the regional depots, the local offices, the field engineers cars, and at the customer site.

8.3. BOARD FLOAT

The predominant service strategy employed today is board swapping. Companies with a decentralized service office and a centralized board repair philosophy have a major problem with this strategy, mainly the enormous PCB float required to enable rapid response to customer calls for service.

Figure 8.2 illustrates a typical repair pipeline. First a customer reports a system down. The remote service center then localizes the problem and sends the customer a good replacement board. The customer swaps boards and his system is up and running. However, the defective board sits at the customer site for one or two weeks before it is shipped to the local service center. In the alternative scenario, field engineers respond to a customer service call and boards are swapped to return the customer's system to operational status. The field engineer in turn, takes the swapped boards and deposits them in the trunk of his car. One to two weeks later, he has accumulated a large number of defective boards and these are turned into the local office for spare parts inventory replenishment. Thirty to 50 percent of these boards will be found good. The local office holds the defective

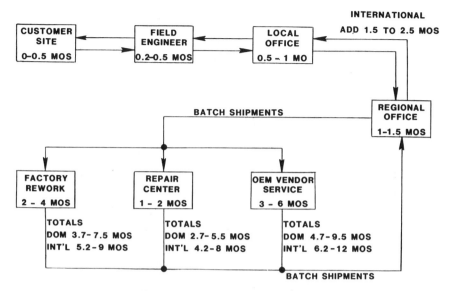

Fig. 8.2. Typical repair pipeline.

boards for 2–4 weeks while sufficient defective spare boards are accumulated to warrant shipment to the regional office. At the regional office, some of the defective boards are fixed. Others are collected until a sufficient number is accumulated to warrant a batch shipment to the factory repair center, or an OEM vendor service.

Typical turnaround times are 2–4 months for factory rework, one to two months for repair center, and 3–6 months for an OEM vendor. The time it takes for a customer's defective board to return to service inventory ranges from 3 to 10 months.

For international customer defective boards, the return-to-service time ranges from 4 to 12 months. Needless to say, the size and lapse time of the board float is staggering. To compound this, a customer's product may go down when a good spare board is unavailable, requiring service to order another spare board on a priority basis from manufacturing. This adds more spare parts into the service inventory, disrupts manufacturing scheduling, and may affect monthly revenue. The size of the spare parts inventory is a function of the vendor's installed base, geographical distribution, product maturity, product

TABLE 8.1. Depot PCB Repair.

	Digital	Analog	Hybrid
Annual PCBs	11,850	3,320	2,830
Batch size	122	56	82
Different PCBs	163	65	122
50% of PCB volume	32	14	49
Tested good	3,555	1,129	821
Average turnaround	4 wks	4 wks	5 wks

mean time between failures, stocking location, stocking spares mix, board float time, and repair turnaround time.

Most service operations have attacked this problem by establishing a captive service spare parts inventory and organizing a network of repair facilities employing repair by local depot, regional depot, third party, and OEM. Table 8.1 illustrates a year's typical depot PCB repair activity. Of 18,000 boards repaired, 66 percent are digital, 18 percent analog, and 16 percent hybrid. The average batch size received from the local feeder offices was 260 − 47 percent digital, 22 percent analog, and 31 percent hybrid.

This depot service center repaired 350 different PCBs. Digital boards accounted for 46 percent, 19 percent were analog, and 35 percent were hybrid. Twenty-seven percent of the 350 different type boards represent an annual volume greater than 50 percent of the total boards repaired. Of the 18,000 defective boards received, 31 percent tested good and were returned to service inventory. The distribution of tested good boards was 38 percent for digital, 33 percent for analog, and 29 percent for hybrid.

The average turnaround time for this service depot is 4 weeks for digital boards and analog boards, and 5 weeks for hybrid boards − considerably less than the 3–10 months discussed previously.

8.4. BOARD MIX

The board mix depends on the industry covered by the service organization. Table 8.2 represents a 1983 estimate of board mix by industry. If the service organization covers computers, one would anticipate the PCB breakdown to be 55 percent digital, 22 percent analog, and 23 percent hybrid. For an industrial service organization, the board

TABLE 8.2. Industry PCB Mix (in percent).

Industry	Digital	Analog	Hybrid
Computer	55	22	23
Peripheral	38	30	32
Office/business	42	30	28
Instrumentation	35	31	34
Communications	37	34	29
Aerospace	36	33	31
Military	35	31	34
Consumer	33	37	30
Industrial	36	34	30
Medical	34	36	30

mix would be 36 percent digital, 34 percent analog, and 30 percent hybrid. However, this is only part of the story. It is desirable for each service organization to stock the proper mix of spare PCBs as a function of use. High-failure PCBs should be stocked in larger numbers than low-failure PCBs — all with a goal of three to four inventory turns per year rather than spare PCBs gathering dust on a shelf. It appears that experience in servicing a product is the only foolproof method to properly balance service stock PCB mix.

8.5. FAULT SPECTRUM

The service fault distribution is significantly different than that found in the manufacturing fault distribution. Every board was functional upon customer acceptance.

Prime Data conducted an extensive survey in 1983 regarding the causes of board failures, and as expected, the fault spectrum varies with the industry. Table 8.3 illustrates this fault distribution. The dominant failure involves digital ICs in all the industries except consumer and medical. In these industries, the dominant failure is transistors and diodes. Depending on the industry, the minority failure is either passive components, interactive components, or transistors and diodes. A summary of the fault spectrum is shown in Fig. 8.3.

The dominant digital IC failure consists of SSI/MSI failures ranging from 18 to 22 percent, plus LSI/VLSI failures ranging from 16 to 21 percent. The consensus is the majority of these failures is caused

TABLE 8.3. Causes of PCB Failures.

Industry	Shorts and Opens	Passive Components	SSI/ MSI	LSI/ VLSI	Transistor Diodes	Interactive Components	Other
Computer	17	11	22	21	14	11	4
Peripheral	18	10	22	19	12	14	3
Office/business	20	14	19	24	12	16	5
Instrumentation	18	11	22	18	21	9	6
Communications	19	12	19	22	14	11	7
Consumer	28	13	21	5	32	3	6
Aerospace	9	4	13	30	25	10	4
Military	10	9	26	27	6	15	4
Industrial	15	12	22	19	17	15	5
Medical	18	11	16	16	21	15	3
Other	13	10	24	16	17	13	5

either by device infant mortality, or end of life, with a small percentage inflicted by service engineer probing of the PCB. The next major failure is transistors and diodes, ranging from 14 to 20 percent. Quality and device stress appears to be the most common cause for these failures.

Shorts/opens encompass the next major failures category, ranging from 13 to 18 percent. Shorts/opens are induced into the PCB by

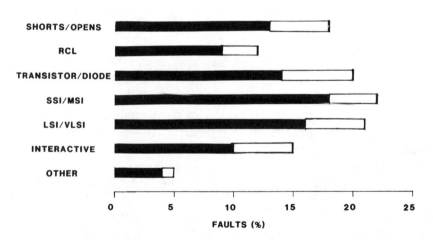

Fig. 8.3. Service fault spectrum.

installing engineering change orders, service engineer probing, and foreign material on the PCB.

Interactive component failures range from 10 to 15 percent. These failures consist of timing delays, input impedance shifts, deterioration of output voltage levels, and distortion because of changes in capacity and internal physics. Also included among these criteria are intermittent failures.

The next failure in order of magnitude is resistors, capacitors, and inductors, ranging from 9 to 12 percent. Unfortunately, there is no data base to assume a failure mechanism.

The last failure category, other, ranges from 4 to 5 percent and is mostly mechanical failures such as broken devices, defective heat sinks, or cracked PCBs.

8.6. PCB POPULATION

The PCB component population distribution is shown in Table 8.4. Forty-eight percent of the PCBs contain less than 100 components, 4,740 of them digital, 2,291 analog, and 1,670 typewritten.

The next largest single PCB category is 3,372 digital PCBs containing between 601 and 1200 digital ICs. The largest populated PCB, of which 33 are repaired annually, has more than 2400 analog components. An understanding of the PCB mix allows service management to select the proper ATE, optimizing test strategies both in factory repair and deployment to the field in depot or local repair centers. Because digital ICs are the dominant fault, service management must have the proper test equipment to effectively resolve this issue. Table 8.5 shows IC technology. SSI and MSI is mostly TTL while LSI, VLSI, and custom ICs are mostly CMOS. This IC mix is 40

TABLE 8.4. PCB Component Population (in percent).

Components	Analog	Digital	Hybrid
<100	69	40	59
101 to 250	6	13	25
251 to 600	4	11	6
601 to 1200	13	32	5
1201 to 2400	7	4	5
>2400	1	0	0

TABLE 8.5. Digital Device Technology (in percent).

Device	TTL	CMOS	ECL	MIX
SSI	60	32	8	40
MSI	67	33	0	31
LSI	40	51	9	16
VLSI	38	49	13	7
Custom	33	51	16	6

TABLE 8.6. Pins per Device (in percent).

Device	1–24	25–40	41–100	101–200	>200
Digital	55	30	10	3	2
Analog	86	9	4	1	0
Custom	29	28	30	9	4

9/83 Internal Survey — 473 Customers.

percent SSI, 31 percent MSI, 16 percent LSI, 7 percent VLSI, and 6 percent custom ICs. The majority of dip, flap pack, SIP, and LCC packages, as illustrated in Table 8.6, have 40 or fewer pins. However, the packaging revolution is strongly leaning toward 41 to 128 pins, causing vendors of service in-circuit testers to aggressively develop new test device interfaces.

8.7. TEST REQUIREMENTS

Service is concerned with testing and repairing a large variety of products: future, current, and discontinued. Table 8.7 shows the distribution of PCB types in 1983, and the projected distribution in

TABLE 8.7. Service PCB Types (in percent).

	1983	1985	Variance
VLSI	11.5	14.0	21.7
LSI	16.8	13.6	(19.0)
SSI/MSI	20.2	17.4	(13.9)
Up	27.2	34.3	26.1
Analog	16.1	12.6	(21.7)
Other	8.2	8.1	(1.2)

TABLE 8.8. Service Cost Breakdown (in percent).

	1983 Expenditure	1984 Budget
Test equipment	7	12
Travel	9	9
Materials	14	13
Salaries	41	38
Burden	13	13
Other	16	15

1985. The increase in VLSI and microprocessor PCBs, along with the decrease in analog boards, illustrates the demanding flexibility the service organization must have both in technical resources and automatic test equipment. By deploying ATE in the field, service organizations can respond to the increasing requirements of repairing high technology products as well as satisfying customer response demands. As an added benefit, this can be accomplished with medium-skilled technical personnel while maintaining a cost-effective business.

Table 8.8 shows 1983 expenditures and 1984 budget for a large east coast service organization. Salaries are the largest percentage, test equipment the smallest. Traditionally, service organizations have been labor intensive rather than capital intensive.

TABLE 8.9. Capital Expenditure.

On–Site		
<$5K	$5K To $10K	>$10K
77%	9%	13%
Center		
<$30K	$30K To $40K	>$40K
45%	44%	9%
Depot		
<$40K	$40K To $50K	>$50K
27%	60%	11%
Factory		
<$50K	$50K To $60K	>$60K
12%	72%	15%

8.8. CAPITAL EXPENDITURE

Based upon a Prime Data survey, Table 8.9 shows the acceptable capital expenditure for service equipment.

The field service engineer carries the test equipment on site. Seventy-seven percent costs less than $1000, while only 13 percent costs more than $10,000.   The present test methodology in the field uses manual instrumentation such as scopes, DMM, and logic analyzers.

For the repair center, 45 percent of the equipment costs less than $30,000, while 9 percent of the equipment costs more than $40,000. The repair center's test equipment consists of programmable IEEE instrumentation controlled by calculators or microprocessor-based controllers, functional testers, in-circuit testers, and in-house-built dedicated testers for specific products or product lines.

At the regional depot repair center, 66 percent of the equipment costs between $40,000 and $50,000.   The depot repair test equipment consists of functional and in-circuit testers, digital logic analyzers, and dedicated test systems.

At factory repair, 72 percent costs between $50,000 and $60,000. The factory repair test equipment is the same type as the depot with the addition of a programming station.   This illustrates the fact that the higher-cost systems are factory-based and as the test equipment is deployed to the field and increases in quantity, the cost of equipment decreases exponentially.

8.9. SERVICE ATE REQUIREMENTS

Based upon service requirements for an extremely flexible test system at a reasonable cost, the criteria employed for selecting field service test equipment are given in Table 8.10.

The initial system price is a function of the budget allotted. This has been discussed previously.   The requirement for an extremely flexible and capable system is evident because of the large and diverse testing requirements placed upon a service organization. Obviously, the service test system should be reliable. Every customer wishes to procure test equipment from vendors with experience and a very sound reputation in a specific industry. In many instances, service is

TABLE 8.10. Service-Dominant Buying Factors.

- Initial system price
- System flexibility/capability
- System reliability
- Vendor's experience/reputation
- Vendor's support
- Programming and fixutre cost
- Programming and fixture time
- Compatibility between factory and field
- Maintenance cost
- Vendor's full product line

required to test products with available test equipment, therefore relying heavily on the vendor's application support. With the large number of different boards service is required to test, the cost and time of both test fixturing and programming is critical.

Deploying in-the-field automatic test equipment compatible with the factory test systems may relieve a large part of the test programming and fixturing burden. The cost of maintenance is always one criterion assessed when selecting a vendor. Also, a customer has leverage over a vendor with product line depth. Many users wish to have the same vendor's product in manufacturing and depot repair, repair centers, and on-site.

Deploying ATE in the depot and service repair centers replaces obsolete testing methods, improves quality, gains custtomer goodwill, minimizes internal support costs, lowers the cost of ownership, and reduces board float.

The pressing question is, then, what test strategies should be employed in company service organizations? Possibilities include use of functional testers, in-circuit testers, digital log analyzers, in-circuit emulators with digital signature analysis, analog signature analyzers, and/or a combination of testers.

8.10. FUNCTIONAL AND IN-CIRCUIT COMPARISON

With the rapid growth of the microprocessor-based boards and the high cost and lengthy time required to generate a microprocessor library model, service has been searching for a more cost-effective method of functionally testing these PCBs. For the past few years, in-circuit microprocessor emulation has been gaining popularity as an acceptable solution. Briefly, in-circuit microprocessor emulation is accomplished by removing the microprocessor from the board and placing it into a micropod. The UUT test program either is developed by a generic algorithm or blowing a PROM. Using a generic algorithm, the in-circuit emulator software generates generic instructions transmitted to the specific microprocessor micropod. Within the micropod, these instructions are translated into the microprocessor instructions code providing total microprocessor control. The PCB is exercised through the microprocessor. The resulting test data are stored in a micropod RAM, translated, and sent to the generic computer for a pass/fail decision. The functionality of the board outside the kernel is tested employing signature analysis accomplished by probing individual nodes. The PROM blowing method entails generating the UUT test program microprocessor address in firmware.

Service in-circuit testers do not require fixtures. Access to the device-under-test is by clip probe, and power is supplied to the board with easy hooks, eliminating the test fixture requirement. The digital test program is developed by calling up the IC-under-test library subroutine, modifying it as a function of how the IC is configured in the circuit, then selecting digital guard points. In many instances, the test program employed in production may be downloaded to the service in-circuit tester. The service in-circuit tester has menu-driven, user-friendly software, in-test program preparation, and operator prompts for test execution with rudimentary failure message and data logging capability.

Table 8.11 presents a condensed comparison between a service functional tester employing in-circuit microprocessor emulation and a service digital in-circuit tester.

Functional testing using in-circuit microprocessor emulation is restricted to 4-, 8-, and 16-bit microprocessor-based boards.

The service digital in-circuit tester tests digital boards by isolating the individual ICs from the surrounding circuitry, and performing logic tests.

In testing A/D and D/A, in-circuit emulators have no analog capability. Digital in-circuit testers perform a gross test while the analog input/output are confined to the programmable logic voltage ranges.

TABLE 8.11. Emulation/In-Circuit Comparison.

Topic	In-Circuit Emulation	Digital In-Circuit
Board types	16-bit microprocessor based	digital
A/D and D/A	none	gross test
Different microprocessor based	no	yes
Two microprocessor based	most cases no	yes
Test methodology	performance	individual devices
Test rate	microprocessor rate	digital – 1 MHz
Go/no-go test time	medium/fast	medium
Diagnostic time	slow	fast
Fault isolation	nodal	component
Fault criteria		
Digital	nodal signature	IC pin fault
Analog	none	none
Fault accuracy	medium	high
Fault detection capability	single	multiple
Fault resolution	nodal/nodal net	component/trace
Failure message	general	specific
Test program		
Methodology	algorithm	library element
Simulation	none	none
Self-learn	node	ROM
Auto-learn	kernel	none
Diagnostics	auto/manual	automatic
Guided probe	difficult	automatic
Debug	low/high	low
Generation time	1–3 mos.	1–2 wks
Quality of test	50–92%	85–94%
Confidence rating	variable	high
ECOs	high	low
Fixturing	micropod	IC chip
Test Execution		
Logic vectors	none	prime method
Signature analysis	4–16 bits	8–bit
Operation	inactive probe	inactive clip
Technician levels		
Test operator	medium	low
Programming	high	low
Maintenance	low	medium
Capital cost	low	medium
Fixturing cost	medium	very low
Major expense	technician	the system
Cost of ownership	medium	low

When two different microprocessors are interconnected on the same board, an in-circuit emulator is incapable of functional testing. The digital in-circuit tester tests the different microprocessors as individual digital elements. When two identical microprocessors are on the same board, depending upon the architecture, the in-circuit emulator may be capable of separating the kernels. Again, the in-circuit tester tests the logic elements as separate entities.

The in-circuit emulator test rate is the microprocessor's normal operating rate. The test rate of the digital in-circuit tester is 1–3 MHz. Go/no-go testing using an in-circuit emulator is quite fast within the kernel. But once outside the kernel, operator intervention is required to probe all the output nodes to ensure proper operation. The service in-circuit tester requires operator intervention to clip each individual IC. This operation averages six ICs per minute.

Diagnostic time using an in-circuit emulator is slow because the operator must probe back, employing documentation from the faulty node to detect the actual faulty node. The go/no-go test time and diagnostic time for an in-circuit tester is the same. In-circuit emulation fault isolation is to the node or node net whereas in-circuit tester fault isolation is to the component.

The in-circuit emulation criterion for a failure is the signature at a node compared with documentation or a signature dictionary. Digital in-circuit tester failure criteria are stuck-at-1, stuck-at-0, or shorted IC pins. Neither the emulator nor the in-circuit tester have analog fault criteria. The fault accuracy for an in-circuit emulation outside the kernel is a function of the programmer's capability and reception of the board architecture to signature analysis, whereas the fault accuracy for an in-circuit tester is dependent upon the library models.

An in-circuit emulator has single-fault detection capability. Each fault must be corrected before remaining faults may be detected. An in-circuit tester has multiple fault capability because each IC is tested individually. All faulty ICs will be detected in a single pass. The fault resolution for an in-circuit tester is specific to the IC, whereas the in-circuit emulator is general, to a node or a series of nodes. Because of the fault detection methodology, the in-circuit emulator's failure message is general and requires interpretation, whereas the in-circuit tester has a specific IC failure message.

The test program methodology for an in-circuit emulator is based upon an algorithm within the kernel and the programmer's knowledge outside the kernel. The test program methodology for the service in-circuit tester is based upon the individual IC subroutines or library elements and the programmer's ability to modify these library models to match the circuit configuration and position digital guard points required to isolate IC outputs. Neither the in-circuit emulator nor the digital in-circuit tester have simulation aides. However, both have self-learn capabilities. The in-circuit emulator operator probes the individual node and learns the signature. The in-circuit tester learns by clipping ROM, PROM, or PAL.

The auto-learn capability of the in-circuit emulator within the kernel is a function of the generic software. The service digital in-circuit tester has no auto-learn capability.

For an in-circuit emulator, the diagnostics within the kernel are automatic, unless microprocessor operation is impeded. Once outside the kernel, manual interaction probing must be employed. To date, there are no in-circuit emulator guided-probe algorithms available. Guided-probe instructions must be manually documented by the programmer. The service digital in-circuit tester has displayed guided clip prompting automatically developed when the individual IC subroutines are merged to form the PCB test programs.

Debugging in-circuit emulator test programs within the kernel is fairly easy. However, once outside the kernel, debugging is difficult, and requires in-depth board knowledge. On the other hand, the debugging of an in-circuit test program involves either modifying a library element or repositioning a digital guard point. The time to generate and debug a test program for an in-circuit emulator ranges from 1 to 3 months. Conversely, a service in-circuit tester programming and debugging rate averages six ICs per hour. This would equate to 1–2 weeks for a complex board. The quality of test ranges from 50 percent to 92 percent for in-circuit emulation, the test quality being a function of the amount of circuitry outside the kernel and/or the kernel algorithm's availability. The service digital in-circuit tester's quality of test ranges from 85 to 94 percent.

Test confidence for an in-circuit emulator varies with the specific board. Test confidence of the digital in-circuit tester is high, a function of the availability of the library elements. Engineering changes to the test program could be medium to high for an in-circuit emulator, but for an in-circuit tester, a library model enhancement generally is all that is required.

The in-circuit emulator fixturing required to test a PCB includes the micropod and power supply easy hooks. For the service in-circuit tester, fixturing consists of power supply and digital guard easy hooks plus an IC clip.

Test execution for an in-circuit tester is accomplished through truth-table or functional-test-vector stimulus and response for SSI, MSI, and LSI logic devices and an 8-bit signature analysis for memories. For an in-circuit emulator, functional tests are performed, employing a microprocessor as a source and signature analysis to measure the response. The test operator of an in-circuit emulator is required to have medium-level technical skills because he must understand the function of the board and signature analysis. The operator on a digital in-circuit tester walks the clip from IC to IC and reads the failure message. Test programming an in-circuit emulator requires a fairly high technical level, generally a test engineer. Test programming an in-circuit tester requires an average technician with an understanding of logic. The maintenance for the in-circuit emulator is very low when compared to the in-circuit tester.

The capital equipment cost for an in-circuit emulator ranges from $5,000 to $15,000, whereas an in-circuit tester ranges from $25,000 to $55,000. The cost of fixturing an in-circuit emulator's micropod ranges from $1,000 to $2,000, whereas the individual service in-circuit IC clips range from $25 to $100. Technical labor comprises the majority of expense in employing an in-circuit emulator as a test strategy. The major expense for in-circuit testing is the system itself. Over a five-year period, the cost of ownership for an in-circuit emulator is 3–4 times that of an in-circuit tester.

The service in-circuit tester in this comparison employs an algorithm to test boards with test vector generation capability. There are digital in-circuit comparison testers on the market with clip access to the IC-under-test while operating within the customer's system. The test circuitry configures this IC in parallel with an IC located within the service in-circuit test system. The user equipment-generated vectors are inputs to both the IC-under-test and the referenced IC. The outputs of both ICs are compared. This test methodology, in many instances, is valid, but a safety question arises when the IC-under-test can be stimulated by the user's equipment. When the IC-under-test outputs are on a bus or clip, access is not hazardous.

8.11. FUNCTIONAL AND IN-CIRCUIT SUMMARY

Both in-circuit emulators and in-circuit testers have advantages and disadvantages. A good test strategy would employ the strengths of both test methodologies.

In summary, benefits of the in-circuit emulation functional tester include:

1. Tests component interaction
2. Tests UUT performance at operational speed
3. Detects critical logic timing faults
4. Lowers initial cost

The service digital in-circuit tester benefits include:

1. Fast test program generation
2. Low technical skill level for test program generation and test
3. Multiple fault diagnostic capability in a single pass
4. Zero fixturing cost

The disadvantages of an in-circuit emulator include:

1. Greater time and technical skill level required for test program generation
2. Restriction to single-base microprocessor boards
3. Low fault-isolation capability requiring greater diagnostic time
4. High technical level required for test program generation and test execution
5. Restriction to available micropods
6. High cost of ownership

The disadvantages of a service in-circuit tester includes:

1. Inability to detect critical logic timing faults
2. Inability to test circuitry interaction
3. Generally does not test at normal operating speed
4. High initial system cost

8.12. GENERIC SERVICE IN-CIRCUIT TESTER

The generic service in-circuit tester is either a desk-size or bench-top configuration. A simplified block diagram of this tester is illustrated in Fig. 8.4. The generic service in-circuit tester consists of a mini-computer or microcomputer controller with 256K bytes of memory. The mass storage unit contains either two dual-density, double-sided floppy diskettes, or a 20 megabyte Winchester disk. The transfer media also is two dual-density, double-sided diskettes. In most cases, the mass storage medium and the transfer medium are one and the same. The video display terminal is a standard, 80-character by

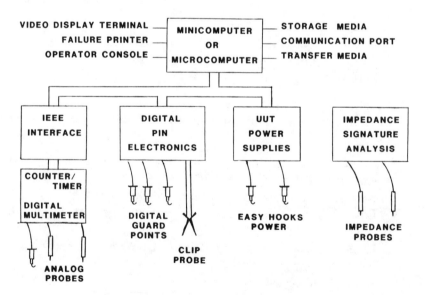

Fig. 8.4. Service in-circuit tester.

24-line display with an alphanumeric keyboard. The failure printer is a small, movable, bench-top, thermal or impact strip printer. The operator's console is a small, movable, bench-top control box with test, continue, and halt buttons and pass/fail status lamps. The system's communication port is typically an RS232 interface. The UUT power supplies typically are fixed, plus and minus 5 volts at either 15 or 35 amps and a programmable 0–60 volts at 15 amps, plus provisions for one or two optional power supplies. The UUT interface to both power and ground is by easy hooks, or edge connector. Most systems offer an IEEE–488 interface with the ability to interface two to three instruments.

The digital pin electronics consist either of sequential static, parallel static RAM-backed pins, or high speed RAM-backed parallel pins. The driver/sensor cards contain 8, 16, or 32 channels per card. The drive is a programmable dual family of ±14 volts sinking or sourcing 500 milliamperes at 30 millivolts resolution and an accuracy of ±10 millivolts. The sensor is a programmable dual-family threshold of ±14 volts at 40 millivolts resolution. The driver/sensor programmability includes: pulse width, pin delay, test frequency, and sensor strobe point. The pin electronics chassis is capable of handling a minimum of 40 pins, generally expandable to a maximum of 128 pins. The digital pin electronics is interfaced with the UUT by a series of clip probes fitting the geometry of various IC packages including dip, flat pack, solder side contact, or conformal coating piercing. In addition, the digital pin electronics contain four to eight digital guard channels. These may be easy-hooked to the UUT.

Analog measurement of the UUT generally is a separate unit employing impedance signature in the form of a scope display pattern or ohmic value.

The service in-circuit tester software emphasis is on simplicity in both test program generation and test execution. Generally, test program generation uses menu-driven software with a response in high-level engineering language. Associated with test program generation is a multitude of programming aids to simplify the technician's task of modifying the individual library elements to conform with the circuit configuration. Test execution is by operator prompting with diagnostics from a specific IC failure to a window logic analyzer display.

9
IN-CIRCUIT TESTER EVALUATION

The preceding chapters have outlined in-circuit testing under categories which are common throughout the industry. This chapter focuses on considerations in the selection of a vendor, beginning with steps to take in-house before entering the marketplace. It is impossible to make a valid decision on a given system without a clear idea of present or projected needs; once requirements have been clarified and priorities assigned, financial models and a vendor evaluation plan can be drawn up. Manufacturers who are thus prepared will be in a better position to compare potential vendors. Specific areas of vendor evaluation are the company itself, software and hardware, fixtures, and product support.

9.1. PREPARATION FOR VENDOR INVESTIGATION

Table 9.1 addresses the steps that a prospective owner of an in-circuit tester should take in investigating production automatic test equipment.

The first is to define the minimum technical requirements that must be satisfied to test the product. In addition, one should define specific needs to relieve real problems, such as what impediments must be resolved in order to have a successful automatic test equipment

TABLE 9.1. Preparation for Vendor Investigation.

- Define your minimum technical requirements to relieve your real problems.
- Define a representative printed circuit board as a standard for vendor comparison.
- Define and assign weighted values to desirable features.
- Define and assign weighted values to intangible requirements.
- Generate a financial model of your present testing method.
- Generate a financial model of a potential in-circuit test system installation. Test your assigned weighted values in terms of cost to the company.
- Establish an evaluation plan including milestones and implementation dates.

program. One must keep in mind what is desirable, what is essential, and what is realistic. The difference usually turns out to be price.

Second is to define a representative printed circuit board that will be used as a standard to compare the vendors. Each vendor will then focus his technical presentation and system demonstration on testing that type of printed circuit board.

This is not to imply that one should have a benchmark. A benchmark is typically a strategic move to slow down a sale, made by a vendor who is losing. To be done properly, a benchmark requires the artwork to be submitted to a vendor for fixturing, which typically takes three to four weeks. A company's representative would go to the vendor with the schematics or parts lists and observe how the parts are coded into the system, to what extent the automatic test generator operates and what warning or not-tested statements result. Next, the company representatives would analyze the debug process to determine how involved it is and what debugging aids are available. The final step would be to integrate this program into a production-level environment by testing 30–80 boards with known failures.

At the end of four to five man-weeks of benchmark evaluation, the question invariably asked is whether the time spent was necessary. Ninety-eight percent of the time the answer is no. Each vendor can satisfactorily illustrate from finished programs all levels required to bring the product into a test environment. Further, each vendor can accurately estimate what can and cannot be tested on a board, how long the fixture and test program preparation will take, and the required test time for the representative board.

The third preparatory step is to define and weight desirable requirements – features that are not absolutely essential but would be nice to have. The weighting factors should reflect a company's technical resources, three-year build plan, and the relative cost to the company.

Intangible requirements such as on-site applications support, service, independent fixture and test programming houses, communication with the vendor, etc., should also be considered. The weighting philosophy should be the same as before.

The next step is to determine what one can afford, beginning with a financial model based on one's present test methods. It is extremely important to precisely determine actual present costs in order to demonstrate that savings will result from adding an in-circuit tester. Similarly, if a new production line is planned, one needs accurate cost projections to make a cost-avoidance justification for including in-circuit testing. Finally, one must generate a financial model of a potential test method employing an in-circuit test system. This will provide a double check or confirmation of the weightings that have been assigned to desirable and intangible requirements. Usually vendors will be happy to provide cost data that will help in developing this financial model.

At this point one should have an understanding of how much money one can afford for a given payback period and return on investment.

Next, an evaluation plan should be established, including milestones to indicate what is going to be accomplished and when. The tendency to go open-loop or manage by interrupts is very tempting when dealing with multiple vendors.

The next step is selecting an evaluation team. Experience has proven that three people is the optimum. However, some companies, in order to prevent hard feelings, want everyone involved in testing to have some voice in vendor selection. If this is the case, a steering committee of three individuals along with a larger evaluation committee is recommended. Selection by committee is an ordeal. Everyone is a steward of the company's money and has his own perception of how it can best be spent. Generally, the optimum selection for an evaluation team is the Manager of Manufacturing and/or Quality Assurance, a Test Engineer, and a Manufacturing Engineer. These three individuals possess the necessary insight into the financial, technical and production arenas.

It is not recommended that a design engineer be a member of the evaluation committee because design engineers think 3–5 years ahead of production. By then the in-circuit tester will have paid for itself two- or threefold. Specifying unnecessary system requirements at added cost and complexity now for things which may be required 3–5 years from now really does not make sense since there may be one or two new generations of in-circuit testers on the market by that time.

The last, and a very important point in the evaluation preparation, is to establish a philosophy on how to treat the vendors. Each vendor should be treated in the same way. One should establish either an open-door policy or a closed-door policy. A closed-door policy usually means assigning one person to two vendors, to talk to those vendors only. There is a tendency for vendors to want to cover all the bases and leverage themselves wherever possible. Committee member disagreements over the merits of different systems should be expected, but disagreements stimulated by vendor disruption should be avoided.

When the selection is reduced to two vendors, plan to assign the same three people to both vendors. They should visit both vendor's plants for a marketing presentation and demonstration of the system, after which an internal decision will be made. Both vendors should be invited in for a final summary meeting before the final decision is made.

At the final summary meeting, it should be announced that on the following Friday afternoon, a final vendor selection will be made with no room for vendor appeal. If the losing vendor requests a debriefing, it is considered professional to accommodate him.

9.2. COMPANY EVALUATION

Table 9.2 outlines eight key issues regarding the vendor's company. Purchasing production ATE is in a sense a marriage between the customer and vendor; therefore, one should be interested in the reputation of the company in the marketplace, both the positive and negative. Every company does have one or two skeletons in its closet and a prospective customer should not be influenced by one skeleton but rather by the company's overall reputation.

TABLE 9.2. Company Evaluation.

- Reputation
- Financial Position
- Dedication to Product
- System Reliability
- Resources
- Market Share
- Support Policy

Second is the financial position of the company as to its capability of support and development of its systems. A customer does not want to help finance his own system with up-front money but to acquire the standard terms of net 30 days. The paramount question is, "Will the company be in the same business one year from now?"

The vendor's dedication to the product is significant in that it gives some insight to the support, future system development, and whether he really has the desire to remain in the in-circuit business. A build rate in excess of 12 hybrid in-circuit testers per month is a good indication of a vendor's dedication.

In order to gain some insight to the system's reliability, ask the vendor for the names of three users who have had a system for more than 18 months. Also, obtain the vendor's Quality Assurance Manager's name to find out what their vendor quality programs are, i.e., if they participate in design reviews, what their in-process audits or inspection functions are, what the final system burn-in criteria and quality tests are, and to whom the QA Manager reports. Remember, one factor determining your product throughput is the test system reliability, which is a function of manufacturing quality.

Another issue concerns the resources that are available to support the product in terms of engineering and applications assistance. What resources does the vendor have?

Consider how successful the vendor has been in the competitive arena with his product. An acceptable measurement is his percentage of the market share; a secondary unit of measurement is the vendor's installed base.

The next issue is one of support policy. What are the details of the support philosophy that the vendor has from the time of installation through warranty and the post-warranty period?

A very significant point is accessibility of the vendor — the communication link between the user's company and the vendor's company for supplying accurate information on delivery and support services. Remember that the user of ATE buys more than just hardware and software; he buys solutions to problems which are otherwise difficult or impossible to solve.

TABLE 9.3. System Software: Test Programs.

1. Automatic program generation
   - Input data coding
   - Self-learn capabilities
   - Automatic dataset generator
   - Emulated device parameters
   - Analog/digital guarding limitations
   - Interpretive or compiled language
   - Editing capabilities
   - Warning statements
   - Program efficiency
2. Device library
   - Library listing
   - Type of modeling
   - Ease of modifying library elements
   - Tools to create new device models
   - Availability of contract modeling
3. Test program code
4. Auxiliary listings
5. Debug and verification routines
6. Incorporating engineering changes
7. IEEE-488 general purpose utilities
8. Contract programming services
9. Documentation, application notes, and software updates

9.3. TEST PROGRAM SOFTWARE

System software is divided into two categories. The first is the test program software, illustrated in Table 9.3. The second, the system operating software, will be addressed later.

Table 9.3 breaks the test program software into nine categories, the first of which is automatic program generation. Is the coding from the schematic or the parts list into the computer-aided program generator near-English or is a separate code necessary? Is it possible to type in "R1," "10K," "5%," "POINTS 1 AND 4," or must programmers learn an alphanumeric code with scaling such as "R001," "0018," "050," "X111," "MTZ," etc? The level of the input data code determines the technical level required to perform this function and directly affects the number of errors in programs.

Next on the list is a self-learn capability. This involves putting a loaded printed circuit board on a fixture, hitting the start button, and having the tester generate the entire program by interrogating the board. This capability is common for shorts and continuities testing. No tester today can perform a complete and unattended self-learn function.

Automatic dataset generation is often available. Again a loaded board is placed on a fixture, and the operator is directed to place a clip or probe at various places to learn the nodal assignments. This prevents incorrect nodal information from being passed to the automatic program generator and from there to the test program, a process that can substantially increase program debug time.

What device parameters are emulated by the system? To what tests is each component subjected in an in-circuit environment? On a transistor, does the system measure three or five different tests? On a diode, two or three? An increased number of tests performed on each component increases the likelihood that good and bad parts will be correctly identified.

What are the analog and digital guarding limitations? In the automatic program generating algorithm, what is the sphere of guarding, how many nodes away from the device to be measured can be guarded, and how many guards can be provided simultaneously? In the digital guarding arena, how many ICs can be preset to known states and how are bus-oriented devices handled?

The next point that is fairly significant is whether the runtime code produced is interpretive or compiled. This is an indication of the flexibility of test generation and an indication of the amount of time required to generate a test program. If the language is interpretive, there is an interaction; when the program is completed, it does not have to be compiled into assembly language. It should be noted that an assembly or compiled language will run faster in the test system than an interpretive language; therefore, some vendors have the digital library elements in compiled code and the rest of the test program in an interpretive language. In general, assembly-type language is harder to edit and debug.

How is the resultant program edited? Is the editor a character or line editor, or does it contain some macro-instructions so that blocks of data can be moved around?

Warning statements from the automatic program generator identify what was not tested and why. In some instances the software gives directions as to how to manually attack the device in question.

The last point under automatic program generation is program efficiency, which relates to the percentage of faults detected. The test system should offer some way of telling the operator how much of the actual board is being tested.

The second major category of test program software is the device library. A customer should request a library listing so that he knows what models currently exist. A customer should also consider what type of modeling is employed. In the digital world, is the modeling by truth table, functional tables, signature analysis, cyclic redundancy check, or some other type? In addition, what analog modeling is available and in what form?

Another concern with the device library is the ease of modifying library elements. When a digital IC is installed on a board, the universal model has to be modified according to the interconnections. Is this modification a major or minor task and what aids are available?

The next point concerns the available tools to create new device models. Does the simulator have simple building block elements in the library such as gates and flip-flops; does it have the flexibility to combine present library elements to create a new device; or does it have a combination or some sort of descriptive LSI language? The main question is, can a customer create his own library elements or do they have to be done by the vendor?

The last point under device library is the availability of contract modeling. In addition to the vendor as a contracting source, are there independent programming houses that will perform this service?

The third main category of test program software is the test program code. Are there various tests and are they readily readable? Is the code English-like or are the statements in assembly language, requiring interpretation? Can the test engineer readily understand what test is being performed at a specific time to aid him in debugging or analyzing a particular test?

The fourth main division of test program software is auxiliary listing. As a byproduct of the test program software generation, the auxiliary listings consist of parts lists, wiring lists, interdevice connections, feedback loops, etc. Are these listings readily available under software control?

Debugging and verification routines are really aids in debugging the test program and verifying optimum performance. Typically, a known-good-board can be placed on the test system and a verification routine can be called up to adjust the test program sequence steps, swap the sense and force leads and remove the guards (as the test requires) while searching for optimum repeatability and fast test time.

How easy is it to incorporate engineering changes? Is this accomplished by amending the present test program or does it require a patch, branch or a link? Does the test program have to be recompiled before board testing can continue?

9.4. OPERATING SYSTEM SOFTWARE

The system operating software is divided into three categories: (1) the operating executive, which is a function of the CPU architecture and instruction; (2) the test executive, which converts the CPU into the system controller; and (3) the utilities that operate under the test executive. Table 9.4 amplifies these three categories.

TABLE 9.4.   System Software: Operating System.

Operating executive
- Type of language
- Overlay, blocked, or linked storage medium construction
- Foreground/background capability
- CAD/CAM data interface ability
- Monitoring variables with flags
- Interface with host networks
- Peripheral drives

Test executive
- Portable executive software
- Test program controls
- Failure message and display
- Automatic wait time
- IEEE-488 instrument controls
- Board-under-test contact verifications
- SA, CRC, TC, truth table, or functional digital stimulus/measurement

Utilities
- Editors — files, macros, characters
- Data logging and analysis capabilities
- Automatic test program loading
- Self-check and fault diagnostics

Documentation, warranty and software updates

Under the operating executive we are interested in the type of language that is required to talk to the CPU, be it FORTRAN, BASIC, or other. Second is the storage medium construction and whether the data are overlaid, blocked, or linked. Third is whether the CPU is capable of running foreground/background, which means testing a PCB and at the same time programming a different PCB in the background mode. Fourth is the capability of incorporating CAD/CAM data as an aid in developing the dataset for automatic program generation. The fifth item concerns the variables which are monitored internally and what flags are raised. Sixth is the ability to interface with host networks; that is, the capability of up- and downloading to a host computer. The last topic concerns the computer peripherals which may be added to the system.

The second main category under operating software is the test executive. A prime question today is whether the test executive software is portable; that is, whether the CPU is essentially invisible for common architecture. This could be accomplished by writing the executive software in a high-level language, such as PASCAL. To date, no vendor offers a portable operating test executive.

Another question is, "What are the test program controls?" These are such commands as "SINGLE STEP," "HALT ON FAIL-URE," "CONTINUE," "PAUSE," etc. Next is the failure message and display. When a failure occurs, what does the failure display field consist of? Is the display field coded, such as "R1004(+)," which means that R4 failed high?" Is the failure message which is printed out the same as the one that is displayed?

The fourth point under the test executive is automatic wait time. This is the capability of the software to strobe a voltage reading at fixed intervals until it is stabilized and then take a measurement. The advantage here is the speed of testing RC-type circuitry and the ability to properly identify a wrong part even if the RC constant changes. In many test systems, to allow time for capacitors to charge, fixed wait-time periods have to be inserted during debug to compensate for RC time constants. However, in this instance, if the installed capacitor is much larger than called for in the circuit, the resistor would also fail. Automatic wait time is therefore an excellent feature if throughput and repeatability are important.

Fifth is the capability of controlling external instrumentation via the IEEE-488 bus. What software drivers are included in the test executive? How are instrument software drivers developed? Does the test executive have a general-purpose IEEE software driver where one could build his own instrument controls by adding the appropriate address and listen/talk codes for a specific instrument?

Sixth is the contact verification routine for the board-under-test. Before the test sequence begins, the test system runs a probe/board continuity test to ensure that the probes are making contact with the board. This software package eliminates false errors due to inappropriate contact, that is, failing good PCB parts because of poor contact.

Finally, in testing digital logic, does the test executive include signature analysis, cyclic redundancy check, transition counter, truth tables or functional digital tables for stimulating and measuring the devices under test, or is a combination of the stimulus/measurement offered?

The third main point under system operating software concerns the utilities that are available. The most important utility is the editor. Is there an option of character, page, file, or macro-instruction editing? Is there the capability of logging and analyzing failure data? Can the test program be automatically loaded, which means that when a fixture is placed on the tester the appropriate program is loaded from mass storage ready to be executed. Very essential are the self-check and fault diagnostics of the test system. This is the capability of putting a calibration adapter on the test system and verifying the accuracy and specifications of the measurement and sense sections of the test system by executing a comprehensive calibration routine. Fault diagnostics localize a malfunction in the test system to a particular component or components.

The last main point involves the documentation, warranty, and software updates which are available with the test system.

9.5. TEST SYSTEM HARDWARE

The third main category of consideration is the test system hardware shown in Table 9.5. The overall system block diagram will exhibit the test system philosophy. Note how the system controller talks to the various elements of the tester. Evaluate the test system's

TABLE 9.5. Test System Hardware.

System architecture
Controller memory and mass storage
Standard and optional Peripherals
Foreground/background
Operator's console
IEEE-488 controller
Digital pin controller
Digital drive/sense voltages and currents
Digital test pattern rate
Digital test rate
Digital testing methodology
Analog matrix
Analog measurement specifications
Maximum number of digital and analog pins and trade-offs
Maximum number of hybrid pins
Total system cost per hybrid pin
Cost of additional pins
MTBF and MTTR
Hardware debug aids
Standalone programming station
Documentation, training, and warranty

architecture in terms of reliability, flexibility, and expandability. The modularity of the system gives some visible insurance against obsolescence. The two main portions of the test system are the controller and the pin electronics. Is the controller a minicomputer or a microcomputer and what are the I/O limitations? What are the controller's memory and capacity? Is the storage medium floppy disk, hard disk or magnetic tape? From a reliability standpoint, core memory is optimum but speed is sacrificed. The hard disk is considered best for mass storage.

What standard and optional peripherals are offered with the test system? Does the tester support foreground and background operation? What is the hardware required to incorporate these options?

Is the operator's console convenient, simple and easy to operate? Is the console designed to enable the operator to perform the minimum amount of action with valid performance?

Does the system have an IEEE-488 controller and how many instruments will it accommodate? What is required to add an additional instrument to the test system?

Hybrid testers have separate digital and analog capabilities; if both the analog and digital capabilities are combined on one pin, the pin is considered universal or hybrid. Many systems have limitations regarding hybrid pins. For example, a tester might have the capability of 1000 digital pins and 500 analog pins. However, when a pin is made hybrid, the limitations might be 400 hybrid plus 500 digital pins. So the actual pin count of the machine decreases.

Typically, the digital driver/sensor architecture routing through the analog matrix determines the capability of the hybrid pins. The first question is, "Do the digital pins have an independent controller?" This would mean that the test pattern vectors are passed to a microprocessor controller which controls the actual digital driver/sensor voltages and pins. Second is, "What is the driver/sensor voltage range and current capability?" The digital source/sink current is extremely important in in-circuit testing, and there should be at least 400 mA capability for driver and 350 mA capability for sink.

The next main areas of concern are the digital pattern rate and the digital test rate. It is essential to distinguish between these two terms. The digital pattern rate is the rate at which patterns can be supplied to the boards under test. This may be the clock serial input data. Frequencies of 100–300 kHz are common. This masks the usable criterion, which is the reciprocal or pin change rate. The time required for a digital pin to be switched from a low to a high and back to a low is typically 1–5 microseconds. This rate, in an in-circuit environment, is typically in the 50–275 kHz range.

Next is the digital testing methodology employed. Is it CRC, signature analysis, transition count, truth table, functional testing or a combination? Many test systems have the capability of CRC, transition counting, and truth tables. Others have a signature analysis and functional testing. It is important to understand how the digital logic is being tested and the flexibility of the hardware.

The analog matrix is essentially a big relay switch that connects the various UUT pins to the analog measurement unit. The system's flexibility is a function of the number of switching points per card and the number of cards that may be installed in the system, plus the bandwidth of the matrix. The type of analog matrix incorporated in the system is typically defined by the number of wires per switching point. Two wires per point is ideal. Unfortunately, on the PCB the analog parts are interconnected and must be electrically isolated in order to be measured. This is accomplished by employing the op-amp and/or bridge technology of nulling currents in the analog measurement unit. The point where the current is to be effective zero is called the guard. Therefore, a three-wire matrix is a force, a measure, and a guard; a four-wire matrix is a force, a measure, a guard, and a guard sense; a six-wire matrix is a force plus sense, a measure plus sense, and a guard plus sense. A Kelvin connection may be accomplished by double pinning a three-wire matrix. Each type of matrix has its advantages and disadvantages depending on the type of PCB being tested.

The next point concerns the analog measurement specifications. In today's in-circuit environment the analog specifications are similar to the instrument specifications. For instance, one vendor's specification for a 10–10K resistor is 1 percent plus 5 ohms, and another vendor's specification is 0.9 percent plus 32 ohms. Closer examination reveals that the positive variance is 20 ohms and the 1 percent system has higher accuracy than the 0.9 percent system. Vendors are not trying to be devious but do try to obtain an advantage whenever possible. This does not imply that one should merely match the individual specifications, but one should clearly understand that the majority of components that are being measured are in the 5–20 percent range. However, one important specification is the frequencies at which impedances are measured. The higher the frequency, the better the part identification.

9.6. VENDOR SUPPORT

The final major issue in selecting an ATE vendor is support (Table 9.6). First under support is the field service policy. When the system fails, how long will it be before the system will be back on the line? Some vendors have a 24-hour service policy, some have a 6-hour policy, and some have a variable policy as a function of the service contract. The next obvious question is, "Where are the field service people located and what is their availability?" How many field service engineers are there per system installation? What is their technical support backup and how many service contracts is each field service engineer responsible for? Once a field service person

TABLE 9.6. Vendor Support.

Field service policy
Availability and location of field engineers
Availability and location of spare parts
On-site applications assistance
Availability and location of applications engineers
Fixture applications engineer
Service contracts
Contract programming services
Contract fixturing services
Contract board test services
Hardware, software, and documentation updates
Test program compatibility
Technical support group depth
Source listing policy
User group

is on the scene and determines the failure, what is the availability of spare parts? Where are the spare parts stocked and to what level? If a part is not in stock, what is the availability?

The next point is on-site applications assistance. Is there an adequate availability of applications engineers in the field and where are they located? How long will it take to respond to a customer's request for assistance? Will the applications engineer satisfy unique fixturing problems? What is the availability and flexibility of service contracts? In a case of heavy production loading or the introduction of a new product, one may be interested in external programming, fixturing and board testing services both from the vendor and from independent houses.

What is the policy for hardware, software, and documentation updates? Is there a standard calendar update schedule or is it on an "as-required" basis?

Test program compatibility is essentially divided into two categories. One is the compatibility of the test programming software from one generation test system to operate on the next generation test system. Second, is the software on System A operating on System B?

A serious concern is the technical support depth. Essentially this is saying, "What is the depth of the bench where you have the main players on the front line, and what happens if the main players are not available?" Another good question is whether the quality assurance department is available for assistance and/or consultation.

Next consider source listing policy. What is the company policy? Can one have a copy of the source listings if he signs a nondisclosure agreement, or are the source listings considered proprietary with no possibility of receiving them?

The last item under support is whether the vendor supports an active "users' group," where customers get together on a periodic basis to hear papers and share ideas and experiences.

9.7. RATIO EVALUATION

Implying ratios as a tool to evaluate a company's performance, by analyzing the company's annual report, is common in the business. This methodology is also common practice in evaluating ATE. As in business there are merit ratios and subjective ratios. The subjective ratios in ATE are referred to as common. The common ratios employed to evaluate automatic test equipment are summarized in Table 9.7. The first common ratio cost/feature(s) could apply to the cost per test point, the cost per kilobyte of memory, or the cost per IEEE instrument. Cost/benefit(s) ratio may express an increase in throughput or productivity. The cost/enhancement ratio may refer to software option networking, test point expansion, etc.

The cost/test program ratio is the total cost of generating board test programs divided by the programming time in hours. The availability ratio is the amount of time that the system is operational minus the time in hours required to repair, divided by the time in hours the system is operational.

The flexibility ratio is the capability of a test system to be reconfigured to test a different product. The flexibility ratio is the total setup time in hours, including test program generation, fixturing, debugging, and production integration, divided by the fault coverage percentage, multiplied by a scaling factor of 1000. The average test time is self-explanatory. The time in minutes required to test a PCB lot is divided by the lot size. The cost/test ratio is the cost of testing a PCB lot in dollars divided by the lot size.

The throughput ratio is a process time, in minutes, divided by the good PCB yield of that process. The productivity ratio is the total

TABLE 9.7. Common Ratios.

Cost/Feature(s)	=	$\dfrac{\text{Cost of Feature(s)}}{\text{Feature(s)}}$
Cost/Benefit(s)	=	$\dfrac{\text{Initial Feature Cost}}{\text{Benefit(s)}}$
Cost/Enhancement	=	$\dfrac{\text{Cost Option(s)}}{\text{Option(s)}}$
Cost/Test Program	=	$\dfrac{\text{Programming Cost}}{\text{Programming Time}}$
Throughput	=	$\dfrac{\text{Total Process Time}}{\text{PCB Lot}}$
Productivity	=	$\dfrac{\text{Total Process Time}}{\text{Finished Units}}$
Cost/Utilization	=	$\dfrac{\text{Operation Cost}}{\text{Usage Time}}$
Cost/Labor	=	$\dfrac{\text{Labor Cost}}{\text{Test Process Time}}$
Availability	=	$\dfrac{\text{MTBF} - \text{MTTR}}{\text{MTBF}}$
Flexibility	=	$\dfrac{\text{Setup Time}}{\text{Fault Coverage Percentage}}$
Average Test Time	=	$\dfrac{\text{Total Test Time}}{\text{PCB Lot}}$
Cost/Test	=	$\dfrac{\text{Testing Cost}}{\text{PCB Lot}}$

manufacturing process time, in minutes, from raw material inventory to finished units divided by the number of finished units. It is common to use the throughput ratio and productivity ratio interchangeably.

The cost/utilization ratio is the cost of operating the test system expressed in an amortized dollars divided by the total time, in minutes, that the system was operated. The cost/labor ratio is the cost of the operator divided by the total test process time, including handling, testing, discretionary, and idle time.

The evaluation ratios of substance, or merit, are expressed in Table 9.8. When considering the setup of a test system to begin

TABLE 9.8. Merit Ratios

Cost/Preparation	=	$\dfrac{\text{Amortized Setup Cost} \times \text{Setup Time}}{\text{Fault Coverage Percentage}}$
Cost/Effectiveness	=	$\dfrac{\text{Recurring Costs}}{\text{Rework Yield}}$
Cost/Efficiency	=	$\dfrac{\text{Recurrint Costs} \times \text{Throughput Time/PCB}}{\text{Number of Finished Units}}$
Cost/Performance	=	$\dfrac{\text{Cost}}{\text{Preparation}} + \dfrac{\text{Cost}}{\text{Effectiveness}} + \dfrac{\text{Cost}}{\text{Efficiency}}$

testing a new product, time and money are critical criteria, along with the flexibility of the test system to test different PCBs as the manufacturing load shifts.   Generally time is more critical than money.   The cost/preparation ratio is the amortized setup cost multiplied by the set-up time, in days, divided by the fault coverage percentage, multiplied by a scaling factor of 1000. The total setup cost consists of the cost to generate the final test program, manufacturing and debugging of the test fixture, test system utilization time, and the miscellaneous aids. The setup cost is generally amortized over the product's number of finished units required to meet product revenue plans for the specific number of years, or simply amortized over a number of years.

Most electronics companies employ cost accounting to derive a standard cost of manufacturing a product, the sum of a materials standard, and a labor standard. Material usage and labor efficiency, workmanship, are the two standard variances that effect production test.

The cost effectiveness ratio is the recurring cost divided by the rework yield expressed in the number of PCBs. This cost effectiveness ratio expresses the cost of the entire test station including test effectiveness and rework effectiveness. The cost effective ratio reflects the cost of obtaining finished units. The cost effective ratio is the recurring costs multiplied by the test time per PCB in minutes divided by the number of finished units.

The widely used definition of cost/performance ratio is the sum of the cost preparation ratio, plus the cost effectiveness ratio, plus the cost efficiency ratio.

TABLE 9.9. In-Circuit Merit Ratios.

Cost/Preparation	=	$\dfrac{\$370 \ \times \ 13.9 \ \text{days}}{90\% \ \times \ 1000}$	= 5.7
Cost/Effectiveness	=	$\dfrac{\$851}{354 \ \text{PCBs}}$	= 2.4
Cost/Efficiency	=	$\dfrac{\$851 \ \times \ 2 \ \text{min}}{956 \ \text{PCBs}}$	= 1.8
Cost/Performance	=	5.7 + 2.4 + 1.8	= 9.9

Table 9.9 illustrates the in-circuit merit ratios employing the data base of our example of 1000 controller boards from the previous chapters. Table 9.10 summarizes the merit ratios for each of the four prime PCB testers, and their in-tandem configurations. In this example, medium volume, medium first-pass-yield as one would suspect, the cost/preparation ratio for a functional board tester is extremely high compared to the cost/performance ratio of the in-circuit type testers. For the in-tandem configurations, the cost/preparation ratios are not additive. The setup costs for each individual tester is added, then divided by the resultant fault coverage. The cost effectiveness ratio of the functional board tester is again higher than the in-circuit type testers. However, in an in-tandem configuration, it is significantly lower than as a standalone unit.

The cost/efficiency ratio of the functional board tester again is higher than the in-circuit-type tester. Significant improvement in the cost effectiveness ratio is obtained when configured in tandem with an in-circuit-type tester. Mathematically the cost/performance ratio is the sum of the other three merit ratios.

The cost/preparation ratio is dependent on the production volume and first-pass-yield. However, the functional board tester's cost efficiency and cost effectiveness ratios would significantly decrease as first-pass-yield increased. Conversely the in-circuit tester's cost efficiency and cost effectiveness ratio will dramatically increase.

TABLE 9.10. Merit Ratio Summary.

Merit Ratio	LBS	ICA	ICT	FBT	LBS-ICT	ICA-ICT	LBS-FBT	ICA-FBT	ICT-FBT	ICA-ICT-FBT
Cost/Preparation	0.2	0.5	5.7	60.3	13.5	9.2	90.4	69.2	99.8	110.8
Cost/Effectiveness	1.5	1.1	2.4	10.5	2.8	2.8	7.6	5.8	7.0	7.3
Cost/Efficiency	0.5	0.5	1.8	31.4	2.6	2.9	14.1	6.5	7.9	7.7
Cost/Performance	2.2	2.1	9.9	102.2	18.9	14.9	112.1	81.54	114.7	125.8

10
FINANCIAL JUSTIFICATION

While purchasing a test system requires careful investigation of vendor, system, and support, the ultimate question is, of course, whether the cost of the system can be justified. Below is a checklist of the most important items which will account for the cost of ownership. Rather than elaborating on them in general terms, this section presents an unaudited case study of a typical installation.

Cost of Ownership Checklist

- Vendor investigation
- System installation
- Test system cost
- Test programming
- Test fixturing

- Operation cost
- Device modeling
- Training
- Maintenance
- System expansion

10.1. PRODUCTION IN-CIRCUIT TESTER

Table 10.1 details the cost of vendor investigation. Tables 10.2 and 10.3 describe a "present test process" with functional testers operating in parallel with repair stations. Tables 10.4–10.6 outline a proposal to integrate in-circuit testing at the site, with the same board volume. The tables demonstrate the time and dollar savings which could reasonably be expected, along with the costs of the system, test programs, fixtures, and special device models. Tables 10.8–10.10 summarize the financial benefits of in-circuit testing, in terms of yearly expenses and savings, and cash flow over a seven-year period.

10.1.1. Vendor Investigation Cost

Table 10.1 summarizes a case study illustrating the activities and manpower a company spent in selecting the proper vendor to solve

TABLE 10.1. Vendor Investigation Cost.

Activities	Hours	Men	Total
Generate a Manufacturing Process Overview	2	4	8
Defining the Test Requirements	3	4	12
Defining the Vendor Evaluation Criteria	3	5	15
Establishing the Vendor Evaluation Package	1½	2	3
Searching for Qualified Vendors (6 Vendors)	1½	3	4½
Vendor's Initial Fact Finding Meeting (4 Vendors)	6	3	18
Internal Briefing Meetings (36 Meetings)	12	4	48
Vendor's Technical Presentations (4 Vendors)	24	3	72
Verification of Vendor's Data (4 Vendors)	4½	3	13½
Evaluating Vendor's Data Meeting (4 Vendors)	10	3	30
Vendor's System Demonstration (3 Vendors)	20	3	60
Evaluating Vendor's Performance (3 Vendors)	4	3	12
Visiting Vendor's Plant (2 Vendors)	16	3	48
Vendor Selection Meeting (2 Vendors)	2½	3	7½
Vendor's Final Summary Meeting (2 Vendors)	4½	3	13½
Generating Vendor Selection Report	4	3	12
Generating the Financial Justification	3½	2	7
Management Presentation Meeting	2	7	14
Vendor's Debriefing Meeting (1 Vendor)	1½	3	4½
Miscellaneous Vendor Meetings (22 Meetings)	5½	2	11
Total Manhours			413½

Loaded Manhour Rate $56.25/Hour

Manpower Cost	$23,259
Travel Expenses	2,855
Miscellaneous Costs	454
Total Vendor Investigation Cost	$26,568

its testing requirements. Of the 413.5 manhours of activities, 227 were spent face to face with vendors; that is, a little better than 45 percent of the time was spent in internal company activities. Typically, the internal activities consume 45–55 percent of the allotted time. The total manhours are multiplied by the burden rate of $56.25 per hour, for a total manpower cost of $23,259. Adding the other direct expenses brings the total vendor investigation cost to $26,568. Too often, this significant expenditure is neglected in the test system cost justification.

TABLE 10.2. Present Test Process.

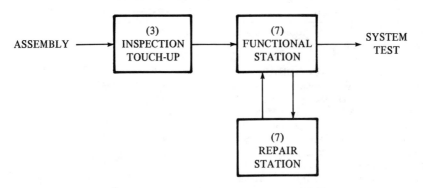

Board Model	Annual Volume	Time in Minutes per Board				Faults per Board	Reject Rate
		Inspection	Test	Analyze	Repair		
101	6,425	3	6	20	6	1.14	45%
102	4,040	4	9	58	7	3.09	68%
103	4,950	3	4	20	5	1.7	65%
104	2,700	5	12	70	6	1.7	66%
105	8,100	3	3	9	4	1.4	52%
106	8,700	4	3	9	3	1.4	57%
107	2,035	12	36	130	16	1.92	55%
108	4,040	5	8	44	10	2.8	66%
109	3,700	7	24	90	6	2.4	60%
110	6,950	3	3	9	4	1.6	48%
Total	51,640	3,571 Hrs	6,546 Hrs	16,357 Hrs	2,772 Hrs	1.8	56%

	Hourly Rate	Burden Rate	Loaded Rate
Inspector (20)	$4.75	225%	$15.44
Technician (20)	$6.25	225%	$20.31

Operate Two Shifts per Day

TABLE 10.3. Present Test Process Expenses.

Board 101 @ 55% Yield (Best Case − 1 Retest)

Inspection	6425 Units ×	3 Min ×	$15.44/60 Min =	$ 4,960	
Test	6425 Units ×	6 Min ×	$15.44/60 Min =	$ 9,920	
Analyze	2891 Units ×	20 Min ×	$20.31/60 Min =	$19,572	
Repair	2891 Units ×	6 Min ×	$15.44/60 Min =	$ 4,464	
Retest	2891 Units ×	6 Min ×	$15.44/60 Min =	$ 4,464	
	Board 101 Present Annual Process Expense				$43,380

Annual Board Volume @ 44% Yield (Best Case − 1 Retest)

Inspection	3,571 Hours ×	$15.44/Hour =	$ 55,136
Test	6,546 Hours ×	$15.44/Hour =	$101,070
Analyze	16,357 Hours ×	$20.31/Hour =	$332,211
Repair	2,772 Hours ×	$15.44/Hour =	$ 42,800
Retest	3,666 Hours ×	$15.44/Hour =	$ 56,603
	Present Annual Process Expense − All Boards		$587,820

10.1.2. Present Test Process

Table 10.2 is the first of nine tables that cover the cost savings which justify a hybrid in-circuit test system for "worst case" conditions. Table 10.2 is based upon the present perspective manufacturing test philosophy. As illustrated by the block diagram, there are three inspection touchup stations feeding seven functional test stations in parallel with seven repair stations. The chart defines 10 board types with their annual volumes plus the amount of time per board to inspect, test, analyze the failure, repair with the average number of faults, and the reject rate. We see that this particular manufacturer has 51,640 boards which require 3,571 hours of inspection time, 6,546 hours of test time, 16,357 hours of analysis time, 2,772 hours of repair time, with an average of 1.8 failures per board at a reject rate of 56 percent.

The time per individual task is converted into dollars by multiplying by the appropriate hourly burden rate. In this instance inspectors cost $15.44 per hour and technicians cost $20.31 per hour. Each individual board, 101 through 110, would then be costed. Illustrated in Table 10.3 is Board 101 at a 55 percent yield rate. Inspection of 6425 units at three minutes per unit, at the loaded rate of $15.44 per hour, divided by 60 minutes per hour, gives us a total inspection cost of $4960. Repeating the same process for testing, our total is $9920. For analysis, the number of units is 6425 times 45 percent failure which equals 2981 units times 20 minutes at $20.31, which is the technician loaded rate divided by 60 minutes, which equals $19,572. These units would then be repaired and retested. The total present annual processing expense would be $43,380 for Board 101. The same exercise would be done for all 10 boards. An accumulation

TABLE 10.4. Proposed Test Process.

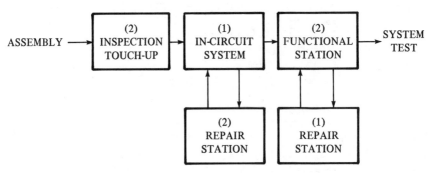

Minutes per Board

Board Model	Annual Volume	Visual Inspection	In-circuit Test	PCB Repair	Functional Test	Functional Analyze	Functional Reject Rate
101	6,425	1	0.29	3	2	5	3%
102	4,040	2	0.43	4	3	15	5%
103	4,950	1	0.32	3	2	5	6%
104	2,700	2	0.57	3	4	18	7%
105	8,100	1	0.38	2	1	2	6%
106	8,700	3	0.31	2	1	2	3%
107	2,035	5	1.27	8	12	33	9%
108	4,040	2	0.34	5	3	11	5%
109	3,700	3	0.75	3	6	23	6%
110	6,950	1	0.38	2	1	2	5%
Total	51,640	1,589 Hrs	365 Hrs	1,477 Hrs	2,136 Hrs	407 Hrs	5%

1. Visual Inspection is mechanical and cosmetic
2. PCB Repair Test is costed as multiple operations
3. 4 seconds load and 5 seconds unload is included in test time
4. Functional Tests are reduced
5. Analyze Tests are reduced to active failures

is shown on the bottom half of Table 10.3 which is the annual board rate at an average of 44 percent yield where the total inspection cost of all the boards is $55,136, the testing cost is $101,070, analysis cost is $332,211, repair cost is $43,800, and the cost of retest is $56,603. Therefore, the annual cost of production testing using the present test method at this facility is $587,820.

10.1.3. Proposed Test Process

Table 10.4 shows the proposed test method, inserting an in-circuit tester and reducing the number of inspection stations from 7 to 2 and the number of repair stations from 7 to 3; two repair stations being in parallel with the in-circuit tester, and one being in parallel with the functional tester. The inspection stations now become cosmetic inspection stations. Taking the same boards and the same volume, we estimate the visual inspection time, the in-circuit test time, the repair time and the functional test and analysis time. For that same Board 101, our visual inspection time is down to one minute, our in-circuit test time is 0.29 minutes, the repair time is three minutes, the functional test time is two minutes, and the functional analysis time is five minutes. The functional test rejection rate is three percent as compared to the previous rate of 45 percent and the overall rejection rate, which is not stated, is less than one percent. Making the same analysis with each individual board, we come up with a total of 1589 hours for visual inspection, 365 hours for in-circuit testing, 1477 hours for repair, 2136 hours for functional testing, 407 hours for functional analysis, with an overall rejection of 5 percent, which is considerably down from the previous rate of 56 percent. This takes into consideration 4 seconds for loading and 5 seconds for unloading each board. The reduction in functional test time is due to the pretesting of the in-circuit tester, and the analysis time is reduced mainly to active-component failures.

Taking Board 101 at its previous yield rate of 55 percent into the in-circuit tester with the assumption that there are going to be two fault-level failures per board producing a yield of 97 percent at functional test, it would take $1653 to inspect the boards. The first pass at in-circuit testing would cost $479 for the toal of 6425 units.

TABLE 10.5. Proposed Test Process Expenses.

Board 101 at a 55% Yield with In-circuit Testing at Two Fault Levels per Board at a 97% Yield Followed by Functional Testing

Inspection	6425 Units × 1	Min × $15.44/60 Min =	$ 1,653
Test, in/c	6425 Units × 0.29 Min × $15.44/60 Min =		$ 479
Repair (45%)	2891 Units × 3	Min × $15.44/60 Min =	$ 2,232
Retest, in/c	2891 Units × 0.29 Min × $15.44/60 Min =		$ 216
Repair (48%)	1388 Units × 3	Min × $15.44/60 Min =	$ 1,072
Retest, in/c	1388 Units × 0.29 Min × $15.44/60 Min =		$ 104
Test, F	6425 Units × 2	Min × $20.31/60 Min =	$ 4,350
Analyze, F	193 Units × 5	Min × $20.31/60 Min =	$ 327
Repair, (3%)	193 Units × 3	Min × $15.44/60 Min =	$ 149
Retest, F	193 Units × 2	Min × $20.31/60 Min =	$ 131

Proposed 101 Board Annual Process Expense $10,713

Projected Annual Board Volume at a 44% Yield with In-circuit Testing at Two Fault Levels per Board at a 95% Yield Followed by Functional Testing

Inspection	1,589 Hours × $15.44/Hour =	$ 24,534
Test, in/c	365 Hours × $15.44/Hour =	$ 5,636
Repair (56%)	1,477 Hours × $15.44/Hour =	$ 22,805
Retest, Inc.	204 Hours × $15.44/Hour =	$ 3,150
Repair (48%)	709 Hours × $15.44/Hour =	$ 10,947
Retest, Inc.	98 Hours × $15.44/Hour =	$ 1,513
Test, F	2,136 Hours × $20.31/Hour =	$ 43,382
Analyze, F	407 Hours × $20.31/Hour =	$ 8,266
Repair, (5%)	74 Hours × $15.44/Hour =	$ 1,143
Retest, F	126 Hours × $20.31/Hour =	$ 2,559

Proposed Annual Board Volume Process Expense $123,935

TABLE 10.6. Capital Requirements and Depreciation.

Capital Requirements	
Hybrid In-circuit Test System	$138,070
Sales Tax of 4%	5,523
Total	$143,593

10% Investment Tax Credit $13,807

Depreciation
Depreciation Method – Straight Line
Estimated Life – 7 Years
Residual Value – 5% or $6,904
95% Amortized Over 7 Years is (95%/7) 100% = 13.6%/Year
$138,070 × 13.6%/Year = $18,788/Year Depreciation

With the predetermined reject rate, 45 percent of the boards, or 2891, require repair. In addition, these boards must be retested on the in-circuit tester. If we assume that 52 percent are short failures, the first fault level of testing, 48 percent of them would be component or other failures on 1,388 units. After repair of these units they would be retested, producing a yield of 97 percent. This would go to the functional tester, which would test all 6425 units. Three percent would fail, 193. These 3 percent would be repaired and retested for a total yield of 99+ percent. The overall cost for the proposed manufacturing procedure would be $10,713, as compared with $43,380 for the old test method. Performing the same type of analysis and modeling on all ten boards, we come up with the data illustrated in the bottom half of Table 10.5 for a total annual board volume processing cost of $123,935 as compared to the initial cost $587,820.

TABLE 10.7. Expenses (Worst Case Fixed Dollars).

A. Fixtures and Test Programs

Board	ICs	Discretes	Point Sets	Fixture	Program
101	6	125	120	$ 717	$ 1,055
102	33	300	390	$ 1,461	$ 3,090
103	33	300	390	$ 1,461	$ 3,090
104	85	406	720	$ 2,427	$ 5,392
105	85	406	720	$ 2,427	$ 5,392
106	6	125	115	$ 703	$ 1,055
107	24	500	410	$ 1,544	$ 4,200
108	32	114	260	$ 1,116	$ 1,758
109	56	149	420	$ 1,572	$ 2,723
110	32	114	260	$ 1,116	$ 1,758
Total Expenses				$14,544	$29,513

B. Special Device Modeling

| Special Device | Time in Engineering Manhours | | | | Total Time |
	Evaluation	Coding	Debug	Documentation	
006-1713	16	8	32	12	68
006-1984	32	16	32	16	96
006-2723	40	20	40	20	120
006-3846	24	16	40	16	96
006-4216	8	8	16	8	40
Total	120	68	160	72	420

Engineering Manhours at $65.00/Hour × 420 Hours = $27,300

10.1.4. In-Circuit Tester's Cost

Table 10.6 shows the estimated capital cost of a hybrid in-circuit tester for this particular application at $138,070, plus sales tax of four percent ($5523), for a total of $143,593. Considering the tax credit available, $13,807, this would be reduced later in the analysis.

Addressing depreciation using the straight line method and an estimated life of 7 years with a residual value of 5 percent, the depreciation rate would be 13.6 percent per year, or $18,788.

10.1.5. Setup Cost

Table 10.7 addresses the additional costs of fixtures, test programs and generating device models for special devices. Table 10.7A addresses each board individually. Board 101, for example has six ICs, 125 discrete components and requires a fixture with 120 probes. The total fixturing cost would be $717 and the test program generation cost for this board will be $1055. The total expense for fixtures is $14,544 and for test programs is $29,513. In addition, there are five special devices which require modeling. Taking the first device in Table 10.7B as an example, the number of manhours required to evaluate device 006-1713 would be 16. After evaluating, there would be eight hours required to code a test for this device, 32 hours to debug it, and 12 hours to supply the documentation, for a total of 68 hours. At a loaded engineering rate of $65 per hour, the total cost for modeling all five special devices would be $27,300.

10.1.6. First Year's Savings

Table 10.8 addresses the expenses or savings per year. The present method costs $587,820. The proposed one costs $123,935, for a savings per year of $463,885. The savings per month would be $38,657, which is really a function of the efficiency level. The first month we assume an efficiency level of 15 percent, the second month 35 percent, the third month 80 percent, the fourth month 90 percent, and finally at the fifth month, we would be 100 percent efficient. Using these percentages as weighting factors, we come up with a savings in the first year of $394,302. The annual expense rate of

TABLE 10.8. Test Process Expense/Savings.

Present Test Process Annual Expense		$587,820
Proposed Test Process Annual Expense		$123,935
Expense Saving per Year		$463,885

Saving per Month is $463,885/12 = $38,657/Month

D + 1 Month	$38,657 × 15% Efficient	=	$ 5,799
D + 2 Months	$38,657 × 35% Efficient	=	$ 13,530
D + 3 Months	$38,657 × 80% Efficient	=	$ 30,926
D + 4 Months	$38,657 × 90% Efficient	=	$ 34,791
D + 5–12 Months	$38,657 × 100% Efficient × 8 =		$309,256
	First Year Savings		$394,302

D = Date of Installation

Annual Expense

Depreciation		13.6%
Interest	13.5%	
Maintenance	8.0%	
Insurance	1.5%	
Property Tax	3.0%	
	39.6%	

39.6 percent is determined by summing the estimated costs for depreciation, interest, maintenance, insurance and property tax.

10.1.7. First Year's Cash Flow

Table 10.9 addresses the first year's cash flow. As determined from the previous table, the first year's savings would be $394,302. The expenses would be the 39.6 percent, determined from the previous table, times the $138,070 capital investment, for $54,676. The fixtures would be $14,544, the test program would be $29,533, the special device modeling would be $27,300, the manufacturing setup costs would be roughly $5500, the recommended spare parts would be $7500, the sales tax would be another $5523, and the fixture, the program and the modeling sales tax would be $2900, for a total first-year expenditure of $147,431. This brings the first year net savings to $246,871. Considering that most corporations are in the 50 percent income tax bracket, we have an expenditure of $123,436

TABLE 10.9. First/Subsequent Year Cash Flow

First Year Cash Flow		
First Year Savings	$394,302	
Annual System Expense at 39.6% × $138,070		$ 54,676
Test Fixtures		$ 14,544
Test Programming		$ 29,533
Special Device Modeling		$ 27,300
Factory Start-Up Costs		$ 5,500
Recommended Spare Parts		$ 7,500
System Sales Tax 4%		$ 5,523
Fixture Programming, Modeling Sales Tax		$ 2,855
First Year Expense		$147,431
Net First Year Savings	$246,871	
50% Income Tax		$123,436
Add Depreciation	$ 18,788	
10% Investment Tax Credit	$ 13,807	
First Year Net Cash Flow	$156,031	

Second and Each Subsequent Year Cash Flow		
Process Saving per Year	$463,885	
Annual System Expense at 39.6% × $138,070		$ 54,676
Additional Fixtures		$ 3,800
Additional Programming		$ 6,500
Additional Device Modeling		$ 4,550
Personnel Training		$ 1,100
Total Year Expense		$ 70,626
Net Annual Saving	$393,259	
50% Income Tax		$196,630
Add Depreciation	$ 18,778	
Net Annual Cash Flow	$215,408	

for taxes and to this we add the first year depreciation of $18,788 plus the 10 percent investment tax credit of $13,807, for a first year net cash flow of $156,031.

10.1.8. Subsequent Years' Cash Flow

The second and subsequent years' cash flow are computed in essentially the same way. The previous savings per year are $463,885. We have the percentage expenditure and we have estimated additional fixturing, program and device modeling at $14,850, and additional personnel training at $1100 for a total second and future year expenditure of $70,626. This would produce a net annual savings of $393,259. Again at a 50 percent tax, the expense would be $196,630 and adding the depreciation, the annual cash flow is $215,408.

10.1.9. Payback and ROI

Table 10.10 demonstrates the cash flow through the first seven years based on a discounted cash flow rate. The payback is then determined by the initial cash outlay divided by the discounted cash flow: $138,070 divided by $135,747 divided by 12, which gives a payback period of 12.2 months. The return on investment after taxes would be the discounted savings of $135,747 divided by the system cost which is $138,070, or 98 percent in the first year. Needless to say, the second, third, fourth, etc., years show a substantial profit.

Every company has its own method of determining payback and return on investment based upon acceptable standard accounting

TABLE 10.10. Cash Flow Summary.

Year	Cash Flow After Taxes	Discount Rate	Discounted Cash Flow
0	$124,263	1.000	($124,263)
1	$156,031	0.870	$135,747
2	$214,929	0.756	$162,486
3	$214,929	0.658	$141,423
4	$214,929	0.572	$122,939
5	$214,929	0.497	$106,820
6	$214,929	0.432	$ 92,849
7	$214,929	0.376	$ 80,813

*In-circuit Test System Cost Less 10% Investment Tax Credit.

When the Investment Cash Outlay is Positive, the Payback is realized.

Payback = Investment Cash Outlay divided by the Discounted Cash Flow divided by 12 Months

Payback = $138,070/($135,747/12 Months) = 122

ROI

$$\text{ROI After Taxes} = \frac{\text{Discounted Savings}}{\text{System Cost}} = \frac{\$135,747}{\$138,070} = 98\%$$

TABLE 10.11. Present Process Return to Service.

	Northeast Depot	
Spare board inventory	$330,000	
Annual repair volume	700,000	
Repair cost at 30%		$210,000
Inventory carrying cost at 20%		140,000
Shipping and handling cost at 3%		21,000
Annual Expense		$371,000

practices. The tables here have not been audited, as the purpose of this case study is to give some conceptual thoughts in developing a layman's cash flow, payback, return on investment and financial model. The detailed financial model will be developed by a company's accounting department. This model is considered worst case. When all the numbers were massaged for this particular case study, the payback period turned out to be 10.8 months with a very good first year return on investment of 140 percent.

10.2. SERVICE IN-CIRCUIT TESTER

The second example of service in-circuit tester justification, is examining the case study of a computer peripheral manufacturer justification to deploy a service in-circuit tester in their northeast depot.

10.2.1. Present Repair Process

Table 10.11 is the first of eight tables that cover the cost savings which justified a digital in-circuit tester for worst case conditions. Table 10.11 shows the northeast depot doing an annual repair volume of $700,000, with a spare board inventory of $330,000. The depot repair cost for defective boards is $210,000 or 30 percent of the annual repair revenue. In addition, the northeast depot incurred an inventory carrying cost of $140,000 plus shipping and handling costs of $21,000 for an annual expense of $371,000, which is 53 percent of their annual revenue.

TABLE 10.12. Proposed Process Return to Service.

	Northeast Depot
Repair cost @ 9%	$ 63,000
Inventory carrying cost @ 15%	105,000
Shipping and handling cost @ 4%	28,000
Annual Expense	$196,000
Inventory reduction @ 15%	$ 49,500

10.2.2. Proposed Repair Process

Table 10.12 shows the northeast depot reducing their cost of repairing defective PCBs to $63,000 by deploying a digital in-circuit tester. In addition the inventory carrying costs are reduced to $105,000 and the shipping and handling costs are reduced to $28,000 for an annual expenditure of $196,000 or 28 percent of their annual revenue, or a cost reduction of 25 percent. Further a reduction in spares inventory of $48,500 or an inventory of $280,000.

10.2.3. In-Circuit Tester Cost

The deployment of the in-circuit tester as shown in Table 10.13 includes a capital investment for the in-circuit tester of $46,000 plus a 4 percent sales tax, as well as an annual expenditure of 26 percent consisting of interest, insurance, maintenance, and property tax, for a total of $11,960. The straight line, seven year depreciation of the in-circuit tester would amount to $6,571 a year. Further there would be a 10 percent investment tax credit of $4,600.

TABLE 10.13. Cost Justification — Deployment of an In-Circuit Tester.

In-circuit tester				$46,000
Sales tax @ 4%				1,840
Annual expense @ 26%				11,960
Interest	15%	Maintenance	5%	
Insurance	2%	Property tax	4%	
Depreciation — SL/7 years				6,571
10% investment tax credit				$ 4,600

TABLE 10.14. First Year's Savings.

	Present annual expense		$371,000
	Proposed annual expense		196,000
	Savings per year		$175,000
D + 1 month	14.6 × 65% efficient	=	9,479
D + 2 months	14.6 × 75% efficient	=	10,937
D + 3 months	14.6 × 85% efficient	=	12,396
D + 4 months	14.6 × 95% efficient	=	13,854
D + 5–12 months	14.6 × 100% efficient × 8	=	116,666
	Reduction in inventory	=	49,500
	First year's savings		$212,832

D = Date of installation

The service in-circuit tester's first year expenses are illustrated in Table 10.14. First is the annual expense of $11,960. Next is the cost of generating 17 board test programs, for a total of $14,760. In order to test the current PCBs, eight new IC subroutines must be generated, for a total cost of $4,500. A service in-circuit tester accesses the board by an IC clip, therefore, the test fixturing cost is zero. The estimated cost for installing the service in-circuit tester in the northeast depot is $1,700, and the spare parts purchased with the system amounted to $11,250. Add to this the system's sales tax at four percent, $1,800. The first year expense is $45,970.

10.2.4. First Year's Savings

The annual savings demonstrated in Table 10.14 amounts to the cost of repairing the PCBs by the present methodology at an annual expense of $371,000 minus the proposed methodology with an annual projected expense of $196,000, resulting in an annual savings of $175,000.

After the system is installed, the user must go through a learning curve before he is 100 percent efficient. The annual savings of $175,000 a year is equivalent to $14,600 per month. During the first month, the user is 65 percent efficient for a savings of $9,479. The second month he is 75 percent efficient, and so on, until the fifth month where he is 100 percent efficient. Therefore the total year's annual savings is $163,332. To this one would add the reduction in inventory of $49,500 for a first year savings of $212,832.

TABLE 10.15. First Year's Cash Flow.

First year's savings	$212,832	
First year's expense		$ 45,970
Net first year's savings	$166,862	
50% income tax		$ 83,431
Add depreciation	$ 6,571	
10% investment tax credit	$ 4,600	
First year's net cash flow	$ 94,602	

10.2.5. First Year's Cash Flow

The first year's cash flow as illustrated in Table 10.15 shows the first year's savings of $212,832 for an expense of $45,970, resulting in a net first year savings of $166,862. Subtracting 50 percent income tax, and adding the system's depreciation and 10 percent investment tax credit, the first year net cash flow is $94,602.

10.2.6. Subsequent Year's Cash Flow

The subsequent year, as illustrated in Table 10.16, shows an annual savings of $175,000 minus an annual expense of 26 percent plus additional test programming, device modeling, and personnel training for a total annual expense of $26,160. The result is a net annual savings of $148,840. Subtracting Federal income tax of 50 percent, and adding the depreciation, the net annual cash flow is $80,991.

TABLE 10.16. Subsequent Year's Cash Flow.

Process annual savings	$175,000	
Annual system expense @ 26%		$ 11,960
Additional:		
• Test programs		7,500
• Device models		4,200
• Test fixtures		–0–
• Personnel training		2,500
Total annual expense		$ 26,160
Net annual savings	$148,840	
50% income tax		$ 74,420
Add depreciation	6,571	
Net annual cash flow	$ 80,991	

TABLE 10.17. Cash Flow Summary.

Years	Discounted Cash Flow	Discount Rate	Cash Flow After Taxes
0	$41,400[a]	1.000	$41,400
1	94,602	0.870	82,304
2	80,991	0.756	61,229
3	80,991	0.658	53,292
4	80,991	0.572	46,327
5	80,991	0.497	40,253
6	80,991	0.432	34,988
7	80,991	0.376	30,453

[a]Testers cost minus 10% investment tax credit

TABLE 10.18. Return on Investment.

$$\text{Payback} = \frac{\text{Cash Outlay} \times 12 \text{ months}}{\text{Discounted Cash Flow}}$$

$$= \frac{\$46,000 \times 12}{\$94,602} = 5.8 \text{ mos}$$

$$\text{ROI} = \frac{\text{Discounted Savings}}{\text{System Cost}}$$

$$= \frac{\$82,304}{\$46,000} = 179\%$$

10.2.7. Payback and ROI

Table 10.17 demonstrates the cash flow through the first seven years based on a discounted cash flow rate. The payback, shown in Table 10.18, is then determined by the initial cash outlay divided by the discounted cash flow: $46,000 times 12, divided by $94,602, equals a payback of 5.8 months. The return on investment is the discounted savings divided by the system cost, or $82,304 divided by $46,000, equaling 179 percent.

REFERENCES

AFIT 3500 Technical Product Description. Fairchild Test Systems. 1983.

Allard, Art. Why in-circuit testing isn't enough. Fluke Automated Systems.

Armstrong, A. L. In-circuit testing of integrated circuits. Faultfinders, Inc.

Bateson, John. In-circuit/functional test system comparison. Fairchild Subassembly Test Systems. 1982.

Bateson, John. Automatic test equipment for printed circuit production. *Insulation/Circuits,* September 1982.

Bateson, John. Production ATE cost/performance comparison. Fairchild Subassembly Test Systems. 1983.

Bateson, John. In-circuit testing of hybrid PCBs – Key issues. Fairchild Subassembly Test Systems.

Blyth, Geoff. Computer-aided repair of PCB. *Evaluation Engineering,* January/February 1982.

Brinton, James. Multiplexing adds pins to board testers. *Electronics,* September 1981.

Carter, Don, and Singleton, Marcel. Overview of Genrad's products for PCB testing. Genrad, January 1982.

Chalkley, Michael. ATE in an engineering environment. Fairchild Test Systems.

Cline, John. The economics of testing evaluation. *Engineering,* September/October 1981.

Cook, Robert, and Lane, Eric. Testing data communication networks in the field. *Electronics Test,* October 1982.

Crosby, Brian. ECL board testing: an in-circuit point of view. *International Test Conference,* 1982.

Friedman, Dan. Understanding and successfully implementing in-circuit testing. Fairchild Test Systems.

Hansen, Peter. Functional and in-circuit testing team up to tackle VLSI in the 80s. *Electronics,* April 1981.

Hotchkiss, Jeff. The economic impact of in-circuit prescreening on functional board test. Teradyne Application Report 125.

Hotchkiss, Jeff. The role of in-circuit and functional board testing in the manufacturing process. *Electronic Packaging and Production,* January 1979.

Hults, Charles, Schwedner, Fred, and Grossman, Stephen. In-circuit test systems – An evaluation. Faultfinders, Inc., April 1975.

Jessen, Kenneth. Overview: Approaches for automatically testing PCBs. *Assembly Engineering,* April 1982.

Jones, Donald. Some common mistakes in industrial product strategy. The Bendix Corporation.

Kennedy, Robert. Service in an automated environment. Digital Equipment Comparison. March 1982.

Lyman, Jerry. Surface mounting alters the PC board scene. *Electronics,* February 1984.

Martin, William. Approaches to automatic testing of electronic assemblies. Zehntel, Inc.

Mechanical Automated Systems. Series 30 Programming Station Product Description. Fairchild Test Systems.

Model 333 Technical Production Description. Fairchild Test Systems.

Model 404 Functional Description. Fairchild Test Systems.

Model 2272 Circuit Board Test System. Technical Description. Genrad, Inc.

Model 3035 Technical Product Description. Fairchild Test Systems.

Model HP3065 System Preliminary Technical Specifications, February 1983. Hewlett-Packard and Company.

Noddin, Fred. Manufacturing data collection systems. Northern Telecom Incorporated. January 1983.

Paul, Eric. An analytical method for measuring the relationship between programming effort and in-circuit and functional test effectiveness. Hewlett-Packard and Company.

Pfaff, Kenneth. In-circuit testing of analog switches and operational amplifiers. Fairchild Test Systems. September 1981.

Prime Data Service Management Information. February 1983. Prime Data, Inc.

Robinson, Rick. Investigation of potential damage resulting from digital in-circuit testing. Hewlett-Packard and Company.

Runyon, Stan. Testing LSI-based boards: Many issues, many answers. *Electronic Design,* March 1979.

Schieber, Steven. Testing issues for PCBs using surface mounted technology. *American Society of Test Engineers Conference,* November 1983.

Schieber, Steven. Ten key considerations in ATE software maintenance. *Evaluation Engineering,* September 1982.

Screen 4400Z Technical Product Description. Fairchild Test Systems. June 1983.

Skilling, J.K. The effect of backdriving stress on digital ICs. Genrad Inc.

Sobotka, Louis. The effects of backdriving digital integrated circuits during in-circuit testing. *1982 IEEE Test Conference.*

Stone, Peter and McDermid, ?. Circuit board testing: cost effective production test and troubleshooting. Hewlett-Packard and Company. March 1979.

Synadinos, Kosmas. In-circuit testing: analysis of empirical information. Western Electric Company.

Test Area Manager Product Description. Fairchild Test Systems.

(Author Unknown) As surface mounting grows more common mixed assemblies rule. *Electronics,* February 1984.

GLOSSARY

Acceptance Tests — A manufacturer's approved test sequence. It demonstrates or verifies the published performance characteristics of a tester. Passing the acceptance tests constitutes the fulfillment of the contractual requirements between buyer and seller. Frequently, rather than the term *acceptance test,* the term *conformance test* is used to avoid the implication of contractual obligation.

Access Time — The time interval between the instant that data is called from or delivered to a storage device (memory) and the instant the requested retrieval or storage reaches the desired location.

Accuracy — (1) The quality of freedom from mistake or error; conformity to truth, a rule, or standard. (2) In testing, the typical closeness of a measurement result to the desired value. (3) The specified amount of error permitted or present in a physical measurement or performance setup.

Acknowledge — A signal that indicates that certain data has been received, or control information has been successfully transferred in a handshaking situation.

Active Device — An electrical element capable of modifying an input voltage in such a way as to achieve rectification, amplification, or switching action, e.g. transistors.

Address Bus — A parallel set of wires over which a digital word is sent to call up an address in memory.

Algorithm — A prescribed set of well-defined rules for the solution of a problem. Algorithms are implemented on a computer by a stored sequence of instructions.

Alphanumeric — Containing both letters and digits, for example, an alphanumeric character set.

Analog Dataset — A list of analog components of a given PCB to be tested for use as input to an automatic program generator. Prepared by a clerk, the analog dataset lists component values, tolerances and nodal points describing electrical circuitry.

Analog Functional Board Testing — The simulation of the PCB environment in the final product by providing various analog test signals stimuli, measuring the board output results under system load conditions, and comparing the results to the stored value for a pass/fail decision.

Analog In-Circuit Testing — Measures individual components on an unpowered loaded PCB. The components are electrically isolated from the surrounding circuitry by guarding, followed by a comparison of the measured value with storage value plus and minus a tolerance, resulting in a pass or fail decision.

Analog Library — For in-circuit testing, a grouping of subroutines, each subroutine consisting of test code for a given analog component.

Apparent Test Points — In a multiplex system, the test points *available* to be connected to the UUT's nodes are *apparent* test points. The multiplex test points that *are* connected to the tester stimulus/measurement unit are *real* test points. Testing is accomplished by switching the group of real test points throughout the apparent test point field.

Archiving — The long-term storage of files on mass-storage media for use at a later time.

Artwork — The original pattern or configuration from which a circuit product PCB is made. Artwork is generated at an enlarged ratio, reduced photographically (to achieve accuracy), then used to produce the PCB. Artwork also is used in this manner to achieve the necessary accuracy during the production of film masks and screens in microelectronics.

ASCII — American Standard Code for Information Interchange. A standard representation for encoding alphanumeric and specific characters as binary values.

Assembler — A computer program that takes symbolic instructions and names of variables, and converts them to a computer-usable program consisting of ones and zeros.

Assembly — A number of parts or subassemblies, or any combination thereof, joined together to perform a specific function.

Assembly Drawing — Document describing the physical structure and layout of a loaded board.

Assembly Language — A programming language in which the programmer can use mnemonic operational codes, labels and names to refer to their numerical equivalents.

Auto-Learn — A software routine which enables the tester to develop a test program from a known-good-board. After entering the UUT's node assignments and location of the individual generic parts, the tester applies the appropriate stimulus to each UUT component and stores the resulting measurement for future test comparison.

Automatic Test Equipment (ATE) — Equipment designed to conduct analysis of functional or static parameters to evaluate the degree of performance degradation, and may be designed to perform fault isolation of unit malfunctions. The decision-making, control, or evaluative functions are conducted with a minimum reliance upon human intervention.

Automatic Test Program Generation (APG or ATPG) — Computer generation of a test program base on the circuit topology, requiring stored library elements and a defined fault algorithm.

Automatic Wait Time (AWT) — A subroutine which automatically inserts a "wait" or "hold" time into a routine until a condition or set of conditions is satisfied.

Average Learn — A software routine which enables the tester to develop an integrated test program by auto-learning a defined number of PCBs and storing the mean average value of each result learned. Some average-learn routines will also derive the individual component's tolerances.

Backdriving — An in-circuit testing technique which drives digital circuitry output pins to a given logic level by supplying pulses of sufficient current magnitude in parallel with the outputs to overdrive the logic state conditions of the next digital device inputs.

BASIC (Beginner's All-Purpose Symbolic Instruction Code) — A popular, general-purpose computer language.

"Bathtub" Curve — A plot of device, component, etc., failures vs. time; used in reliability analysis work. It is so-called because the plot resembles the cross-section of an old-fashioned bathtub. The life-cycle failure rate for virtually all mechanical or electronic devices follows the general shape of this curve.

BDL (Board Description Language) — A high-level language designed for easy understanding. Employing English mnemonics commonly used in testing, the UUT data base file is written in this language.

Bed-of-Nails Test Fixture — A test fixture consisting of a frame and holder containing a field of spring-loaded pins designed to make electrical contact with the assigned internal PCB's nodes. Typically, in-circuit test fixtures are vacuum-activated.

Benchmark — Used to evaluate the performance of testers relative to each other. The benchmark includes set-up, test program generation and fixturing, followed by testing a defined number of a specific PCB. The Benchmark follows the customer's defined procedures developed for particular objective(s) and specific requirements.

Blister — A localized swelling and separation between any of the layers of a laminated base material, or between base material and conductive foil. It is a form of delamination.

Bootstrap — A program used to start up and initialize a computer.

Buffer — (1) A circuit used to restore a signal to a specified drive level, often employed in data transmission through long cables. (2) A temporary software storage area where data resides between time of transfer from external media and time of program-initiated I/O operations.

Buffered Printer — A printer containing its own electronically isolated memory, enabling it to assimilate and print out data at a slower rate than received.

Burn-in — To operate a piece of equipment under power for a period of time to isolate any early-failing parts.

Burn-in, Dynamic — A high-temperature test with device(s) subjected to actual or simulated operating conditions.

Burn-in, Static — A high-temperature test with device(s) subjected to unvarying voltage rather than to operating conditions; either forward or reverse bias.

Burst — The high-speed transfer from memory of a pulse train that starts at a prescribed time and continues for a specified duration (or number of clock cycles).

Byte — The smallest grouping of directly addressable parallel bits from memory, usually 8 or 16, that a computer can process at one time.

Cache — A high-speed intermediate buffer memory usually located between a central processor and the processor's main memory.

Capability — An arbitrary measurement of a test system's performance of a given task(s).

Carriers — Holders for electronic parts and devices designed to facilitate handling during processing, production, imprinting, or testing operations, and to protect such parts under transport.

Central Processor (Also **central processing unit** or **CPU**) — The main control key element of a computer system in which both logical and arithmetic instructions are carried out via program control.

Check Program — A software program designed to check the operation of hardware, or another program.

Checkpoint — A place in a routine where a check, or recording of data for restart purposes, is performed.

Checksum — A character at the end of a block of data that indicates whether the binary sum of all of the characters in the block is odd or even. Used for error detection purposes. Similar to *parity checking.*

Chip Component — An unpackaged circuit element (active or passive) fabricated using one or all of the semiconductor techniques of diffusion, passivation, masking, photoresist, and epitaxial growth for use in hybrid microelectronics. Besides ICs, this element includes diodes, transistors, resistors, and capacitors.

Chip Resistors — Very small chips of ceramic, from 100 mils to as little as 25 mils in length. Chip resistors have extremely low shunt or parasitic capacitance, no inductance, are stable and, generally, low in cost.

Clock — A basic synchronizing signal source within a computer; generates "clocked" signals to maintain correct time and phase relationships.

Closed Loop — A control arrangement in which data from the processor device being controlled is fed to the computer to affect the control operation; that is, the computer can perform all control functions without intervention of an operator.

Cold Solder Connection — A soldered connection where the surfaces being bonded moved relative to one another while the solder solidified, causing an uneven solidification structure possibly containing microcracks.

Comment — A section of a program that has no function other than to explain the meaning of a part of the program; comments are neither translated nor executed, but are simply copied verbatim in the program listing.

Compatibility — Ability of two or more entities to function together without interference. Also implies that no or little modification is necessary when moving a system element (i.e., fixtures, programs) from one system type to another.

Compiler — A software routine which analyzes and converts a test program from high-level test language to binary machine code. The input file is *source code,* the output file is *object code.*

Complex PCB — An arbitrary measurement of a circuit board containing a large variety of different chip types, feedback loops, or a number of VLSI devices.

Component Library − An organized (by model/type, number, etc.) list of components and the associated test routines for each.

Component Side − The side of a printed circuit board on which electronic components are mounted. The other printed circuitry board side is free of components, contains the interconnection artwork, and is called the *solder side.*

Computer − A device capable of accepting data, applying prescribed procedures to the data, and supplying the results of these procedures.

Computer-Guided Probe − A fault-isolation technique based solely on good-circuit data. The probe algorithm acts as the master instruction to an operator to probe various IC pins on the UUT until it derives the final diagnosis and diagnostics.

Computer-Aided Design (CAD) − Mechanical and/or electrical design using a computer to aid in a layout or decision process.

Computer-Aided Manufacturing (CAM) − Manufacturing using a computer to aid in data control and analysis.

Computer-Aided Testing (CAT) − Testing using a computer in a test system (or a central computer linking two or more test systems) to aid in test-data processing.

Conditional End-of-Test − A command in a test program to stop execution of the program when a particular condition or set of conditions is reached.

Contact Bounce − The undesired intermittent closure of open contacts or opening of closed contacts, usually occurring during a change of state.

Continuity Testing − A test procedure wherein voltage is applied to two PCB "tracks" (runs, loads, etc.) that should be interconnected, to determine the presence or absence of current flow. This process is repeated until all interconnections on the board have been tested.

Cost of Ownership − The accumulation of all the recurring expenses associated with a tester over its projected life, expressed on an annual basis. Cost of ownership includes operation labor, tester's maintenance and repair, amortized usage, new test programs and fixtures, and additional personnel training.

Cost of PCB Test − The total expense of testing a group of PCBs. The sum of operator's labor, amortized system usage, amortized setup, and overhead. The test cost per PCB is the total PCB testing expense divided by the number of boards tested.

Crash (System) − The loss of control of a system or computer by getting into an unescapable loop, or by executing an incorrect program.

Current Loop − A closed electrical circuit in which the input and output current are the same (a common value is 20 ma).

Current Spikes − Rapid current changes, often associated with noise or waveform leading-edge distortion.

Data Compression − Technique of delaying a data stream and then compressing (in time) the data to enable analysis and use of the data.

Data Link − Any information channel used for connecting data processing equipment to any input, output, display device, or other data processing equipment, usually at a remote location.

Data Management — Storing, handling, and processing volumes of data.

Datalog Analysis — A program for organizing and mathematically analyzing raw failure data.

Datalogger — A system to measure a number of variables and make a written tabulation and/or record in a form suitable for computer input. In PCB testing, a datalogger accumulates data on PCB failures for future analysis.

Datalogging — Placing into memory the accumulation of raw failure data.

Debug — To examine or test a procedure, routing and/or hardware for the purpose of detecting and correcting errors or flaws. In testing, the process of modifying a test program to obtain the correct results from a series of known-good-boards.

Debug Time — Time required to remove the errors or flaws from a program or piece of hardware.

Debugger — A special program routine which specifically checks for, locates, and flags programming errors.

Density — The number of entities or activities in a given area, expressed as volume or time, or the volume of ICs on a PCB.

Design for Testability — A design process or objective involving a deliberate expenditure of effort to assure that a product may be thoroughly tested with minimum effort, with accessibility to pertinent testpoints, and with high confidence ascribable to test results.

Device — The physical realization of an individual electrical circuit element in a physically independent package which cannot be further reduced or divided without destroying its stated function.

Diagnostic Test — A software subroutine designed to perform fault isolation.

Digital In-Circuit Testing — Testing of digital circuitry by isolating each IC, one at a time, from the surrounding circuitry then driving the inputs with a series of logic test patterns while sensing the outputs for the proper logic response. Conventional fault coverage is generally confined to pin faults. Advanced fault coverage is an abbreviated functional test of the main internal IC's circuit elements.

Digitize — To express data in binary code (digital form). Also, to convert an analog signal to a digital signal.

Digitizing — Any method of reducing feature locations on a flat plane to digital representation of $X-Y$ coordinates. Used in fixture layout and CAD/CAM design.

Dimensional Stability — A measure of dimensional change caused by factors such as temperature, humidity, chemical treatment, age, or stress; usually expressed as units/unit.

Disable — To keep a circuit or component from operating normally by superimposing a fixed or blocking signal on one or more of its input lines.

Discrete Component — An electronic component having an individual identity. Fabricated prior to installation, and/or separately packaged, it is not part of an integrated circuit which is functionally complete by itself, such as transistor, capacitor, or resistor.

Disk Storage — A mass-storage device employing a flat, rotating medium onto which data can be stored via magnetic recording techniques, and retrieved by magnetic playback. The file-structured mass storage device allows fast, random access of data.

Disk Drive — The mechanical system used to rotate and provide means for electrical input/output from a disk memory system.

Diskette, Floppy Disk — A flexible, flat plastic (usually Mylar) disk coated on both sides with a metallic oxide serving as a memory device. It is enclosed in a plastic jacket and looks similar to a 45-rpm record in a jacket.

Down Time — The period during which a system or device is not operating because of internal failures, scheduled shutdown, or servicing.

Driver — (1) A circuit which shapes and amplifies the voltage or current of a waveform for use as an input to a number of other devices, or to a single, high-power device. (2) The digital test circuitry used to force current to overdrive the logic inputs of the UUT.

Driver/Sensor (D/S) — The digital test circuitry used to force logic input levels and monitor logic output levels of the UUT.

Driver/Sensor Controller — A microprocessor controller capable of controlling the digital pin electronics throughout the digital sequence.

Dual-Chamber Fixture — A test fixture with two separate bed-of-nails fields mounted on separate vacuum chambers. They can be actuated separately, and generally are employed as part of in-circuit testing strategy to eliminate the effects of handling time. One PCB is loaded on one test probe field while the other PCB is being tested on the other test probe field.

Dump — To printout or externally store the contents of a computer's memory.

Duty Cycle — The ratio of the sum of all pulse durations to the total period during a specified period of continuous operation.

Early Failures — Also **infant failures** or **burn-in failures**. Failures during the early life-cycle of a device or system, beginning at some stated time and during which the failure rate of some items is decreasing rapidly.

Editor — A program allowing a user to enter data into a formatted computer file, and to manipulate and alter the data. *Editing* is the process of creating or modifying a source file.

Emulation — Simulation of the actions of a device or system in real time. The imitating computer system accents the same data, executes the same programs and, as a goal, achieves the same results as the original device or system.

Enable — To turn on, switch on power, etc. Also, a signal causing an activity to proceed, or a device to produce data outputs.

Endless-Loop Instruction — One which transfers control of itself, thus resulting in its executing indefinitely, or until interrupted by a hardware signal.

Error — Any discrepancy between a computed, observed, or measured quantity and the true, specified, or theoretically correct value or condition.

Failure — (1) (Reliability) The termination of the ability of a component to perform its specified function. (2) (Catastrophic) Both sudden and complete. (3) (Complete) Beyond specified limits, causing total lack of function.

(4) (Critical) Likely to cause injury or significant damage. (5) (Degradation) Both gradual and partial. (6) (Gradual) Could be anticipated by prior examination. (7) (Dependent) Caused by the failure of an associated fault. (8) (Inherent weakness) Subject to stresses within the stated capabilities. (9) (Misuse) Attributable to the application of stress beyond the stated capabilities. (10) (Random) Time of occurrence is unpredictable. (11) (Intermittent) Occurring for a limited time on a random basis. (12) (Nonrelevant) Excluded in interpreting test results. (13) (Transient) Induced by momentary or temporary external factors such as power fluctuation, electromagnetic interference or temperature excursions. (14) (Natural) Malfunction far beyond a projected life.

Failure Analysis — The logical, systematic examination of an item, or the circuitry in which it is used, to identify and analyze the probability, causes, and consequences of potential and real failures.

Failure Data — Information on the nature, frequency, rate, etc., of failures of components.

Failure Rate — The statistical average, expressed as a percent, of the number of failed PCBs occurring in a lot of PCBs. For example, 200 PCBs failing in a lot of 1000 PCBs equals a 20 percent failure rate.

Failure Report — A card, report, listing, etc., containing the details of a failure or failures which have occurred on a board or component.

Fault — (1) (Reliability) A physical condition causing a device, component, or element to fail to perform in a required manner. (2) (Manufacturing) Faults induced in assembly (wrong component, reversed component, missing component, bent leads, open leads, shorted leads) and in the soldering (shorts and opens) process itself. (3) (Bus) A fault in a device connected to a bus resulting in the bus and all other devices connected to it to be held at a fixed state. (4) (Parametric) Any parameter of a device to exhibit a value outside its specified tolerance range. (5) (Short) An abnormal connection of relatively low impedance, whether made accidentally or intentionally, between normally electrically-separated points. (6) (Stuck-At) An IC pin's digital signal being permanently held in one of its binary states. (SAI = stuck at one fault; SA0 - stuck at zero fault). (7) (Timing) Digital device switching occurring to the proper level, but outside of a specified time interval. (8) (Design) A design characteristic of either hardware or software causing or materially contributing to a malfunction. Faults resulting from engineer errors, omissions or oversight during the design phase. (9) (Input) Preventing an input line from responding to the signal normally furnished, usually the result of a shorted or open semiconductor junction or shorted/open bypass capacitor. (10) (Node) Affecting the logic of all pins connected to the node in an identical manner. (11) (Functional or Logic) The failure of an input stimulus to produce the proper output response. (12) (Performance) The failure of a device or group of devices to function in the proper manner because of dynamic or interaction characteristics.

Fault Coverage — An attribute of a test procedure expressed as the percent of faults of the failure population which the test or test procedure will detect.

Fault Detection — A process capable of discovering, or is designed to discover, the existence of faults; the act of discovering the existence of a fault.

Fault Dictionary — A list of elements in which each element consists of a fault signature and all the faults that are detected by each fault signature.

Fault Isolation — When a fault is known to exist, a process identifying, or designed to identify, the location of that fault to within a small number of replaceable components.

Fault Localization — When a fault is known to exist, the process identifying, or designed to identify, the location of that fault within a general area of the circuitry. The process is less specific than *fault isolation*.

Fault Masking — A fault X hides fault Y from being detected.

Fault Modes — The various ways faults may occur, such as stuck faults, bridging faults, intermittent faults, and functional faults.

Fault Resolution — A measure of the capability of a test process to perform failure isolation among replaceable units, generally expressed as a percentage.

Fault Siganture — The characteristic, unique erroneous response produced by a specific fault. It consists of data bits useful for fault detection and isolation in constructing a fault dictionary, each bit representing the existence of a discrepancy between good and faulty response on the output. An output data stream resulting from the test of a unit containing failing test step numbers and failing output bits.

Fault Simulation — A process allowing prediction or observation of a PCB's system behavior in the presence of a specified fault without actually having that fault occur in the circuitry. The process demands modeling of either the fault, or the PCB, or both.

Faults per Board — The statistical average of the number of faults detected divided by the number of PCBs tested. For example, 280 faults detected in testing 1000 PCBs equals 0.3 faults per board.

Faults per Failed PCB — The statistical average of the number of faults detected in a group of defective PCBs. For example, 280 faults detected in a group of 200 defective PCBs equals 1.4 faults per failed PCB.

Fetch — To retrieve an instruction or data from computer memory.

Firmware — Software in hardware form. Encompasses a combination of software and hardware (e.g., computer microcode in ROM).

First-Pass Yield — The statistical average, expressed as a percent, of the number of finished units that pass all production tests without any rework in a lot of manufactured PCBs. For example, a first-pass yield of 60 percent occurs when 600 PCBs in a manufactured lot of 1000 pass all production tests.

Fixture — A device enabling interface of a printed circuit board with a field of spring-contact probes. This device contains either a dedicated head or an interface for interchangeable test heads and a means of keying the product to be tested.

Fixture Kit — Preassembled kit of parts to build a test fixture.

Foreground/Background — A system allowing a computer's memory to be partitioned (divided) and then used independently in either of the two memory partitions.

Glitch — An errant signal introduced at a device's input through unintentional means, causing an undesired output.

Global Command — A command that influences all programs over which it has control.

Go/No-Go Test — A test designed to yield a "test pass" or "go" indication in the absence of faults in a UUT, and a "test fail" or "no-go" indication in the presence of fault(s).

GPIB (General-Purpose Information Bus) — A general multipurpose instrumentation control input/output bus used in test systems.

Gray Code — A binary number notation in which any two numbers whose difference is one are represented by expressions that are the same except in one place or column, differing by only one unit in that place or column.

Ground Loop — Current flow between two or more ground connections where each connection is at a slightly different potential because of the resistance of the common connection.

Guarding — A measurement isolation technique used in in-circuit testing. (1) *Analog guarding* places equal voltages at both ends of a component parallel to the component under test to block the parallel current path. (2) *Digital guarding* disables the output of a device that is connected to the device-under-test output by applying the appropriate input vectors.

Handshake — A transmitting and receiving process in which characters are exchanged to establish synchronization.

Hard Copy — Information printed out on paper, as contrasted to that merely displayed, as on a CRT. Produces a permanent record of the data output.

Hi Rel — High-reliability device. Devices of this nature are designed to extremely tight tolerances, are constructed for long service life, and must meet high-quality control levels.

Initialization — Providing a set of fixed, initial conditions to start a program from a specific point.

Initialize — (1) To establish an initial condition or starting state; for example, to set logic elements in a digital circuit or the contents of a storage location to a known state so that subsequent application of digital test patterns will drive the logical elements to another known state; and (2) to set counters, switches, and addresses to zero or other starting values at the beginning of, or at prescribed points in a computer routine.

Interstrobe Time — The time between a driver strobe and a sensor strobe.

Input/Output — (1) Interface circuits or devices offering access between external circuits and the central processing unit or memory. (2) Pertaining to devices that accept data for transmission to a computer system (input) or accept data from a computer system for transmission to a user or process. Devices that perform both functions are known as *I/O devices* (e.g., *terminals*).

Intelligent Terminal — A VDT capable of responding to user activities with requests or questions, permitting bidirectional flow of information between the VDT and the user.

Interactive — A device (most often a computer peripheral) capable of providing a two-way conversation (dialog) with the user, including questions, prompts, etc.

Jump Instruction — An instruction that places a new flag in the program counter of a computer, in contrast to normal single-step incrementing. The instruction may be "conditional," i.e. it may be placed in the program counter only if certain conditions are met, or "unconditional," in which case it is placed in the program every time, regardless of other circumstances.

Kernel — The "core" of circuitry in a processor or system. The kernel must be functioning properly for the processor or system to successfully execute tests of other portions of itself.

Known-good-board (KGB) — A correctly operating PCB. It is used in learning or debugging a test program in development; and for comparison testers it serves as the standard unit to which others are compared.

Label — A name given to an instruction or statement in a computer program to identify the location in memory of the machine language code or assignment produced from that instruction or statement.

Learning — The process employing software routines in determining output response or a measurement value when an appropriate input stimulus is applied to a component on a known-good-board.

Library — A grouping of subroutines, each subroutine comprising test code for a given component. The library could contain analog or digital device tests.

Listener — A device/circuit which receives data, but does not transmit data.

Loaded Board — A PCB with electronic components mounted on it.

Logic — A mathematical arrangement using symbols to represent relationships and quantities, handled in a microelectronic network of switching circuits, or gates, which perform certain functions; also, the type of gate structure used in part of a data processing system.

Logic Level — At a specific instant the digital voltage level defining a "one" or a "zero."

Looping — Condition in which the output signal is fed back to the input of a circuit; or in software where reaching the end of a routine results in a return to the start of the same routine.

Machine Language — The programming language the computer can directly execute with no translation other than numeric conversion.

Mapped Operating System — An operating system using a common memory divided into two separate "grounds," foreground and background. Routine system operation is performed in one ground (usually foreground), while alternate activities such as program development are done in the background section of memory. This provides time-shared access to memory, which *may* be rapid enough to *appear* as simultaneous operation in both grounds.

Mass Storage — Computer peripheral devices into which large amounts of data can be deposited and recovered. May be referred to as "auxiliary storage" or "secondary storage" to differentiate from memory.

Mean Time between Failure (MTBF) — The arithmetic or statistical mean average time interval (hours) that may be expected between failures of an operating tester. State whether actual, predicted, or calculated.

Mean Time to Failure (MTTF) — Total operation time (hours) divided by total number of failures.

Mean Time to First Failure (MTTFF) — A measurement of reliability giving the expected time before the first failure. Expressed as the statistical mean average time (hours) from the completion of installation and checkout of a system until the first significant failure occurs.

Mean Time to Isolate (MTTI) — The statistical mean average time (hours) that may be expected to achieve fault isolation as measured from the time of fault detection to the time of fault isolation.

Mean Time to Repair (MTTR) — The arithmetic or statistical mean average of time (hours) that may be expected to complete a repair activity.

Microcomputer — A microprocessor complete with stored program memory (ROM), random access memory (RAM), and input/output (I/O) logic. If all functions are on the same chip, this is sometimes called a *microcontroller*. Microcomputers are capable of performing useful work without additional supporting logic.

Microprocessor — An IC package incorporating logic, memory, control, computer, and/or interface circuits, the whole of which is designed to handle certain functions to form a microcomputer that can execute a multitude of different instructions. Each instruction causes output state changes dependent on how its qualifier inputs are programmed and the instruction is put on the input lines.

Modeling — The process of mathematically describing a circuit such that the result from the simulator will correspond to the signal values in the actual circuit.

Multiplexing — The combining of two or more signal channels into one single optical channel by employing an electronic multiposition switch under the control of the test program. Multiplexing reduces the size and cost of a tester's switching matrix while increasing the apparent/real test-point software management.

Multitasking — Executing a number of tasks during a single time period, usually by assigning each slice of time and suspending other tasks waiting for input/output, the completion of the tasks, or external events.

Nested — In programming, a subroutine within a subroutine, or a loop within a loop.

Nodal Pin Assignment — Arbitrary assignment of numbers to test nodes on a schematic diagram or a PCB artwork.

Node — Common connection of two or more components on a schematic diagram or on a PCB. Commonly used as a test point.

Nominal Delay — The average time that signals take to travel through a logic element or along a wire. The effect of an input change to an element on the output will not occur until after the duration of a nominal delay. Usually expressed in nano- or microseconds.

Object Code — Binary machine language output of an assembler code. Object code can be loaded into a system and run as is.

Operating System (also executive routine). — A fixed program that manages the hardware functions and the software files of a system.

Orientation — Positioning or adjusting with respect to a fixed reference; specifically the direction in which an ICs, diodes, and polarized capacitors are placed on a board.

Paper Tape Reader — A device capable of "reading" a paper tape; it converts punched-hole patterns in tape into electrical pulses, usually in ASCII format. In some instances, paper tape also refers to the paper in the test results printer.

Parity Checking — See *Checksum.*

PCB (Printed Circuit Board) — An insulated board containing an overlaid conducting film for connections, on which components are deposited. PCBs are constructed as *single-sided,* interconnection traces on the bottom side, *double-sided,* interconnection traces on the top and bottom sides, and *multilayer,* two or more double-sided boards sandwiched together to form one PCB. The term PCB is used interchangeably when discussing either *loaded,* having components installed or *bare,* free of components, boards.

Peripheral — A device used to extend the operation of a computer, with regards either to functionality or capacity. Peripherals are connected to the central processor of a computer system by appropriate data paths.

Pointer — A register or memory location that contains an address rather than data.

Polling — Determining the states of peripheral or other devices by examining each one in succession.

QA (Quality Assurance) — An organization within a company whose purpose it is to perform ongoing monitoring and assessment of product and procedures in order to assure continued meeting of company quality specifications and objectives.

Race — In synchronous digital logic, the concurrent change of two or more feedback lines. If the final state which the circuit has readied does not depend on the order in which the variables change, then the race is "noncritical." Otherwise it is a critical race.

Race Condition — In logic circuitry, the condition in which one or more logic signals arrive at a device as an improper sequence of events.

RAM-Backed Pin — In in-circuit and functional testers, the digital driver/sensor local memory storage, consisting of input/output data RAM, force or sense enable RAM, masking/ignore response RAM, and error collection RAM, plus possibly data compression RAM memory.

Real Test Points — See *Apparent Test Points.*

Real-Time — A computer process executed with sufficient speed so that the results of a process being monitored appear to be presented instantaneously. The computer generally is able to present the results with sufficient speed to permit control changes to be made.

Real-Time Operating System — One which is designed to interface with programs that have real-time requirements. Also may be referred to as a *real-time executive* (routine) or *real-time monitor.*

Recovery Time — The time required for a signal to return to its rated value after a sudden change (e.g., the time for the output voltage of a power supply to return to its rated voltage after application of a step load).

Receiver — In a system, an electrical panel designed to mate with the contact panel on a test fixture to provide the electrical interface between the system and the board under test.

Redundance, Redundancy — The introduction of auxiliary elements and components into a system to perform the same functions as other elements in the system to improve reliability and safety. Also, the use of additional components, programs, or repeated operations, not normally required by the system to execute its specified tasks, to overcome the effects of failures.

Reject Rate — The number of defective PCBs per number of PCBs tested, expressed as a percentage.

Repair Cycles — The number of times a tester detects a failure on a repaired PCB and returns the defective PCB for rework. A normal repair loop consists of a tester defining a PCB's fault(s), which the repair station corrects and returns for retest or repair verification.

Repair Message (Rework Message) — A message printed out or displayed on a test system, indicating the existence, location, and need for specific repairs, or reworking to the product tested.

Response Time — (1) Time for a device or circuit to respond to a signal input from the instant the signal is injected. (2) The time required for a resulting condition to reach steady state after variation of an input quantity.

Self-Learn — A software routine enabling the tester to develop a test program, without any interaction, directly from a known-good-board. Self-learn is commonly employed in learning shorts/opens test programs.

Schematic Diagram — A drawing showing, by means of graphic symbols, the electrical connections, components, and functions of an electronic circuit.

Simulator — Creates a computer model of an entire digital board to develop patterns of testing and definition of failure conditions. It also supplies test-effectiveness information.

Single Test — Command given to an ATE test system, resulting in running one set of instructions, following which the test operation will stop.

Source Code — A computer program written in assembly language, or in a high-level language.

Spike — A distortion in the form of a pulse waveform of relatively short duration, superimposed on an otherwise regular or desired pulse waveform.

Static Test — A test in which measurement is made of a UUT after, and only after, the outputs have stabilized with respect to a given input stimulus; also, a test made at less than normal system/clock speed (rate).

Step Function — An essentially nonperiodic waveform that has a transition from one voltage level to another, the time of which is negligible compared to the total duration of the waveform.

Stimulus — Any physical or electrical input applied to a device intended to produce a response.

Test Effectiveness — A measure which reflects the fault coverage and fault resolution provided by a test, taken at the next test station.

Test Plan — A documented procedure identifying the PCB to be tested, tests to be performed, setup and test schedule, testers required, resources required, required reports, evaluation criteria, any risks, and a contingency plan.

Test Point — A node within a circuit or system which can be measured or stimulated to facilitate testing.

Test Program Set (TPS) — The total test package for a UUT, consisting of the test program, test fixture, and the instructions for test-program operation.

Test Speed — Length of time to run a particular test or test sequence, transmit the stimulus, monitor the response, and make a pass/fail decision. Stated as *tests per second.*

Testing Accuracy — A measure of the accuracy with which a test measures what it was designed to test.

Throughput — Quantity of acceptable product or finished units turned out during a specified time period.

Timing Diagram — A display of what logic state each of a group of lines is in from moment to moment.

Total-Time-to-Fault Isolation (TTFI) — Time from discovery of a fault to specific isolation of that fault.

Tri-State Testing — Testing involving checking that an electronic device can be maintained in any one of its three stable states.

Truth-Table Testing — Testing by applying electrical signals in accordance with a prescribed logic matrix pattern indicating outputs which should result from a given pattern of inputs.

Turnkey — Generally used to indicate a tester delivery is complete with debugged test program and fixture. Literally means that the customer installs the tester on his production floor and starts testing PCBs.

Uncompromised — An uninterrupted state of events, a steady constant stream of action with no breaking points or alterations.

Unconditional Test — A test without limitation or restriction as to test mode, time, etc.

Unknown State — Most memory elements used in sequential circuits are bi-stable devices. When the power is turned on, they are normally designed to have a more or less symmetrical configuration. The initial states are therefore unpredictable and are called *unknown states.* Unknown states can also be the result of critical races, oscillations, or ambiguity delays.

Wave Soldering — A process wherein printed boards are brought in contact with a gently flowing wave of liquid solder which is circulated by a pump in an appropriately designed solder pot reservoir. The prime functions of the molten wave are to serve as a heat source and heat transfer medium and to supply solder to the joint areas.

Winchester Disk — A high-speed mass-storage disk unit with nonremovable disk.

Yield — The number of acceptable or finished units produced compared to the maximum number possible.

Yield Rate — Quantity of acceptable or finished product turned out during a specific time period (e.g. 2200 boards per month, etc.)

Yield Report — A report providing information on product yield or yield rate.

INDEX

INDEX